THE PLANE WAVE SPECTRUM REPRESENTATION OF ELECTROMAGNETIC FIELDS

IEEE/OUP Series on Electromagnetic Wave Theory

The IEEE/OUP Series on Electromagnetic Wave Theory consists of new titles as well as reprintings and revisions of recognized classics that maintain long-term archival significance in electromagnetic waves and applications.

Books in the Series

Chew, W. C., *Waves and Fields in Inhomogeneous Media*
Christopoulos, C., *The Transmission-Line Modeling Methods: TLM*
Clemmow, P. C., *The Plane Wave Spectrum Representation of Electromagnetic Fields*
Collin, R. E., *Field Theory of Guided Waves,* Second Edition
Dudley, D. G., *Mathematical Foundations for Electromagnetic Theory*
Elliott, R. S., *Electromagnetics: History, Theory, and Applications*
Felsen, L. B. and Marcuvitz, N., *Radiation and Scattering of Waves*
Harrington, R. F., *Field Computation by Moment Methods*
Ishimaru, A., *Wave Propagation and Scattering in Random Media*
Jones, D. S., *Methods in Electromagnetic Wave Propagation*, Second Edition
Lindell, I. V., *Methods for Electromagnetic Field Analysis*
Tai, C. T., *Generalized Vector and Dyadic Analysis: Applied Mathematics in Field Theory*
Tai, C. T., *Dyadic Green Functions in Electromagnetic Theory*, Second Edition
Van Bladel, J., *Singular Electromagnetic Fields and Sources*
Wait, J., *Electromagnetic Waves in Stratified Media*

THE PLANE WAVE SPECTRUM REPRESENTATION OF ELECTROMAGNETIC FIELDS

P. C. CLEMMOW

IEEE Antennas and Propagation Society, *Sponsor*

The Institute of Electrical
and Electronics Engineers, Inc.
New York

Oxford University Press
Oxford, Tokyo, Melbourne

Oxford University Press, Walton Street, Oxford OX2 6DP
Oxford New York
Athens Auckland Bangkok Bombay
Calcutta Cape Town Dar es Salaam Delhi
Florence Hong Kong Istanbul Karachi
Kuala Lumpur Madras Madrid Melbourne
Mexico City Nairobi Paris Singapore
Taipei Tokyo Toronto
and associated companies in
Berlin Ibadan

Oxford is a trade mark of Oxford University Press

IEEE PRESS
445 Hoes Lane, PO Box 1331
Piscatawaty NJ 08855-1331

IEEE Antennas and Propagation Society, *Sponsor*

First issued 1966 Pergamon Press
Reissued with additions 1996 Oxford University Press jointly with IEEE Press

This is a copublication of Oxford University Press in association with IEEE Press

IEEE ISBN 0-7803-3411-6
IEEE Order Number: PC5682
OUP ISBN 0 19 859225 6
Library of Congress Cataloging-in-Publication Data
(data available)

Printed and bound in Great Britain by Bookcraft (Bath) Ltd.

FOREWORD

James R. Wait
Tucson, Arizona

As a commissioning co-editor, with Professor Alex Cullen, I am pleased to see that the IEEE Press and OUP have secured the rights to republish this excellent monograph by Dr Phil Clemmow who had written a long-cherished exposition on the angular spectrum concept. Because the original version published by Pergamon Press did not have any references, an annotated bibliography by Rod Donnelly has been added.

CONTENTS

PART I. THEORY

PART II. APPLICATION

Part I
THEORY

CHAPTER I

PRELIMINARIES

1.1. OBJECTIVE

In the pedagogy of the theory of the electromagnetic field it is customary to pay considerable attention to the plane wave. This is due primarily to the relative simplicity of plane wave solutions of Maxwell's equations; their use enables some of the important elementary physical and engineering characteristics of the electromagnetic field to be elucidated without appeal to other than quite straightforward mathematics.

On the other hand, in somewhat more advanced work, such as diffraction theory, the tradition has mainly been, in the spirit of Huyghens and Fresnel, to think of the field as generated by a distribution of localized sources. Also, of course, the standard retarded potential formulation involving volume integrals over the actual current distribution in effect simply treats each volume element as a dipole.

There is, however, the possibility of continuing to benefit from the simplicity of plane wave solutions by retaining them as the bricks from which to construct whatever more elaborate type of solution arises. This idea has a long history. Its wide exploitation, though, is comparatively recent, as is also the explicit recognition of its close association with the technique of Fourier analysis.

The object of this short book is to explain how general electromagnetic fields can be represented by the superposition of plane waves travelling in divers directions, and to illustrate the way in which this *plane wave spectrum* representation can be put to good use in attacking various characteristic problems belonging to the classical theories of radiation, diffraction and propagation.

It need hardly be said that in a book of this size the problems are not treated exhaustively. To have included alternative theoretical methods, or details of the physical background, or details of the analytical, numerical or physical nature of the solutions, would have tended to swamp the avowed didactic content.

It must also be conceded that various topics that could legitimately be embraced by the title of the book are omitted altogether. Most conspicuous by their absence are problems in which the plane wave spectrum of the field is essentially discrete. Such fields arise typically in cavities and waveguides, and these topics are so fully covered in other books that their inclusion seemed superfluous. Also omitted are problems involving fields in some sense "random" in space or time; their treatment would require the introduction of statistical concepts, which themselves are quite unconnected with the main stream of the mathematical development here presented.

On the positive side of the balance sheet the book offers a largely unified theory of a range of problems, solutions to all of which are obtained in forms at least patently capable of yielding numerical results by straightforward means. The reader is assumed to be competent at integration in the complex plane, but otherwise the discussion is virtually self-contained; the burden of the analysis is carried by the exponential function, and the sprinkling of Bessel functions does not signify the need for any great familiarity with their properties. In this way the aim is to furnish the student of electromagnetic theory with a useful technical tool and a comparatively compact account of some interesting aspects of his discipline.

1.2. MAXWELL'S EQUATIONS

The electromagnetic fields are for the most part assumed to be time-harmonic. The complex representations of the field vectors, with the time factor $\exp(i\omega t)$ understood, are used in the standard way. They satisfy the Maxwell equations

$$\operatorname{curl} \mathbf{E} = -i\omega\mathbf{B}, \tag{1.1}$$

$$\operatorname{curl} \mathbf{H} = i\omega\mathbf{D} + \mathbf{J}. \tag{1.2}$$

In (1.2), $\mathbf{J}$ is the volume current density, and it is associated with the volume charge density ϱ through the charge conservation relation

$$\operatorname{div} \mathbf{J} + i\omega\varrho = 0. \tag{1.3}$$

The divergence of (1.1) gives

$$\operatorname{div} \mathbf{B} = 0; \tag{1.4}$$

and the divergence of (1.2), together with (1.3), gives

$$\operatorname{div} \mathbf{D} = \varrho. \tag{1.5}$$

Any media involved are treated macroscopically, being described by linear constitutive relations between the field vectors, which then denote the "average" fields that would be recorded by conventional laboratory measurements.

If the fields can be regarded as generated by a current in what is otherwise a vacuum, then

$$\mathbf{D} = \varepsilon_0 \mathbf{E}, \quad \mathbf{B} = \mu_0 \mathbf{H}, \tag{1.6}$$

where ε_0 and μ_0 are the vacuum permittivity and permeability. In this case eqns. (1.1) and (1.2) read

$$\operatorname{curl} \mathbf{E} = -i\omega\mu_0 \mathbf{H}, \tag{1.7}$$

$$\operatorname{curl} \mathbf{H} = i\omega\varepsilon_0 \mathbf{E} + \mathbf{J}. \tag{1.8}$$

At points where there is no current density

$$\operatorname{curl} \mathbf{E} = -i\omega\mu_0 \mathbf{H}, \tag{1.9}$$

$$\operatorname{curl} \mathbf{H} = i\omega\varepsilon_0 \mathbf{E}, \tag{1.10}$$

which imply

$$\operatorname{div} \mathbf{H} = \operatorname{div} \mathbf{E} = 0. \tag{1.11}$$

By eliminating one of **E**, **H** from (1.9), (1.10), and using (1.11), it follows that each cartesian component of **E** and **H** satisfies the time-harmonic, homogeneous wave equation

$$\nabla^2 \varphi + k_0^2 \varphi = 0, \tag{1.12}$$

where

$$k_0^2 = \omega^2 \varepsilon_0 \mu_0. \tag{1.13}$$

It is sometimes convenient to appeal to the converse of this last statement, namely that any divergence free vector each of whose cartesian components satisfy (1.12) can legitimately be identified with either **E** or **H** to specify a vacuum electromagnetic field. It is also worth noting that from any solution of (1.9) and (1.10) another can be deduced at once by the transformation

$$\mathbf{E} \to \mathbf{H}, \quad \mathbf{H} \to -\mathbf{E}, \quad \varepsilon_0 \leftrightarrow \mu_0. \tag{1.14}$$

Difficulties associated with the vector character of eqns. (1.7) and (1.8) are significantly eased in the idealized case in which the field is two-dimensional, being independent of one cartesian coordinate, z say. For the equations then separate into two independent groups, namely

$$\frac{\partial E_z}{\partial y} = -i\omega\mu_0 H_x, \quad \frac{\partial E_z}{\partial x} = i\omega\mu_0 H_y,$$

$$\frac{\partial H_y}{\partial x} - \frac{\partial H_x}{\partial y} = i\omega\varepsilon_0 E_z + J_z, \qquad (1.15)$$

and

$$\frac{\partial H_z}{\partial y} = i\omega\varepsilon_0 E_x + J_x, \quad -\frac{\partial H_z}{\partial x} = i\omega\varepsilon_0 E_y + J_y,$$

$$\frac{\partial E_y}{\partial x} - \frac{\partial E_x}{\partial y} = -i\omega\mu_0 H_z; \qquad (1.16)$$

and any two-dimensional field can therefore be regarded as the superposition of an *E-polarized* field, in which E_z, H_x, H_y, and J_z are the only non-zero field components, and an *H-polarized* field in which H_z, E_x, E_y, J_x and J_y are the only non-zero field components. The identification of E_z with any solution of the two-dimensional, time-harmonic, homogeneous wave equation

$$\frac{\partial^2 \varphi}{\partial x^2} + \frac{\partial^2 \varphi}{\partial y^2} + k_0^2 \varphi = 0, \qquad (1.17)$$

completely specifies an E-polarized field in a current free region; the other non-zero field components, H_x and H_y, follow at once from a knowledge of E_z through the first two equations of (1.15). The identification of H_z with any solution of (1.17) likewise specifies an H-polarized field in a current free region.

When isotropic media are considered it is assumed for the sake of simplicity, what is commonly the case in practice, that the permeability differs negligibly from the vacuum permeability μ_0. The constitutive relations are thus taken to be

$$\mathbf{D} = \varepsilon \mathbf{E}, \quad \mathbf{B} = \mu_0 \mathbf{H}, \quad \mathbf{J} = \sigma \mathbf{E}, \qquad (1.18)$$

where ε and σ are the permittivity and conductivity respectively, and $\mathbf{J}$ in (1.18) of course signifies the conduction current. The substitution of (1.18) into (1.1) and (1.2) gives, at points where there is no impressed current source,

$$\text{curl}\, \mathbf{E} = -i\omega\mu_0 \mathbf{H}, \qquad (1.19)$$

$$\text{curl}\, \mathbf{H} = i\omega(\varepsilon - i\sigma/\omega)\, \mathbf{E}; \qquad (1.20)$$

so that the use of the complex representation has the advantage that the effect of conductivity can be readily allowed for by working in terms of the single parameter

$$\varepsilon - i\sigma/\omega, \tag{1.21}$$

which is sometimes called the complex permittivity. The appearance of ω in (1.21) indicates explicitly what may often be the major dependence of the complex permittivity on frequency, but it must not be forgotten that ε and σ are themselves certainly frequency dependent, albeit possibly in effect constant over an appreciable range of frequencies.

Anisotropic media are not treated extensively in this book, but some consideration is given to media that can be characterized, for time-harmonic fields, by the linear constitutive relations

$$\mathbf{D} = \varepsilon_0 \mathscr{K}\mathbf{E}, \quad \mathbf{B} = \mu_0 \mathbf{H}, \tag{1.22}$$

where $\mathscr{K}$ is a tensor. The tensor form of the relation between **D** and **E** means that the two vectors are in general no longer parallel. The relation takes account of all current due to the average motion of the charged particles constituting the medium, in the same sort of way as (1.21), and $\mathscr{K}$ can be frequency dependent and complex. For a lossless medium $\mathscr{K}$ is Hermitian; that is, its i, j element $\varkappa_{ij}$ is identical with the complex conjugate $\varkappa_{ji}^*$ of the j, i element. This result follows from a statement of energy balance, consequent on Maxwell's equations, which deserves brief mention.

The interpretation as power flux density of the Poynting vector $\mathbf{E} \wedge \mathbf{H}$, where **E** and **H** momentarily stand for the actual electric and magnetic fields, is well known. For time-harmonic fields it is commonly only the time-averaged power flux density that is of interest, and this is conveniently obtained in terms of the complex representation of the field from the form $\operatorname{Re} \frac{1}{2}\mathbf{E} \wedge \mathbf{H}^*$. With (1.22), eqns. (1.1) and (1.2) read

$$\operatorname{curl} \mathbf{E} = -i\omega\mu_0 \mathbf{H}, \tag{1.23}$$

$$\operatorname{curl} \mathbf{H} = i\omega\varepsilon_0 \mathscr{K}\mathbf{E}. \tag{1.24}$$

Thus the mathematical identity

$$\operatorname{div}(\mathbf{E} \wedge \mathbf{H}^*) = \mathbf{H}^* . \operatorname{curl} \mathbf{E} - \mathbf{E} . \operatorname{curl} \mathbf{H}^*$$

gives

$$\operatorname{div}(\mathbf{E} \wedge \mathbf{H}^*) = -i\omega\mu_0 \mathbf{H} . \mathbf{H}^* + i\omega\varepsilon_0 (\mathscr{K}^*\mathbf{E}^*) . \mathbf{E}. \tag{1.25}$$

Now in a lossless medium the time-averaged power flux has zero divergence at any point where there is no impressed current source; that is, the real part of (1.25) is zero. The necessary and sufficient condition that this be so is evidently that $(\mathscr{K}^*\mathbf{E}^*).\mathbf{E}$ be real; or, introducing suffix notation and the summation convention, and equating the expression to its complex conjugate, that

$$\varkappa_{ij}^* E_j^* E_i = \varkappa_{ij} E_j E_i^*.$$

If on one side of this relation the dummy suffixes i and j are interchanged, it appears that the condition is indeed $\varkappa_{ji} = \varkappa_{ij}^*$.

1.3. FOURIER INTEGRAL ANALYSIS

There are many ways of expressing the integral representations associated with the names of Fourier and Laplace; these differ in degree of generality, in outlook, in interpretation and in notation. The purpose of this section is merely to record the particular formulation adopted in this book, introducing only those few simple examples that are required subsequently.

The basic concept is the representation of any function $f(\xi)$ of a real variable ξ in the form

$$f(\xi) = \int_{-\infty}^{\infty} F(\eta)\, e^{i\xi\eta}\, d\eta. \tag{1.26}$$

The path of integration is initially presumed to run along the real axis, although distortions permitted by the rules of contour integration may legitimately be introduced later. The spectrum function $F(\eta)$ must therefore at least be defined for effectively all real values of η, and the essence of the Fourier theorem is that for such values

$$F(\eta) = \frac{1}{2\pi} \int_{-\infty}^{\infty} f(\xi)\, e^{-i\eta\xi}\, d\xi. \tag{1.27}$$

The case of paramount importance in the present context is

$$F(\eta) = \frac{1}{2\pi i(\eta - \eta_0)}. \tag{1.28}$$

Suppose, first, that η_0 has a non-zero imaginary part. Then the path of integration in (1.26) can be closed by an infinite semicircle, above the real axis when $\xi > 0$ and below when $\xi < 0$, without altering the value of the integral. The behaviour as $|\eta| \to \infty$ of the exponential factor in the integrand ensures that there is no contribution from the semicircular part of the path, a standard result which it is not difficult to establish rigorously. Once the path has been closed the value of the integral can be written down from Cauchy's residue theorem; then (1.26) gives

$$f(\xi) = \begin{cases} 0 & \text{for } \xi < 0, \\ e^{i\eta_0\xi} & \text{for } \xi > 0, \end{cases} \tag{1.29}$$

when the imaginary part of η_0 is positive; and

$$f(\xi) = \begin{cases} -e^{i\eta_0\xi} & \text{for } \xi < 0, \\ 0 & \text{for } \xi > 0, \end{cases} \tag{1.30}$$

when the imaginary of η_0 is negative. It requires but a trivial direct integration to confirm that the substitution of (1.29) or (1.30) into the inverse formula (1.27) does indeed correctly recover (1.28).

These results need only be expressed in a slightly different way to cater for the case when η_0 is real. It is then necessary to indent the path of integration in (1.26) so that it avoids the pole at η_0. If the path is chosen to pass above η_0, then $f(\xi)$ is given by (1.30); if below, by (1.29). Several immediate deductions from this case are now listed.

By putting $\eta_0 = 0$ it is established that the *unit step function*

$$f(\xi) = \begin{cases} 0 & \text{for } \xi < 0, \\ 1 & \text{for } \xi > 0, \end{cases} \tag{1.31}$$

has spectrum

$$F(\eta) = \frac{1}{2\pi i\eta}, \tag{1.32}$$

with the η path of integration passing below the origin.

A trivial generalization of (1.31), (1.32) is that

$$f(\xi) = \begin{cases} 0 & \text{for } \xi < \xi_0, \\ 1 & \text{for } \xi > \xi_0, \end{cases} \tag{1.33}$$

has spectrum function

$$F(\eta) = \frac{e^{-i\xi_0\eta}}{2\pi i\eta}, \tag{1.34}$$

with the η path of integration passing below the origin.

By subtracting the unit step function (1.33) for which $\xi_0 = a$ from that which for $\xi_0 = -a$ it is established that the *rectangular pulse*

$$f(\xi) = \begin{cases} 0 \text{ for } \xi < -a \\ 1 \text{ for } -a < \xi < a \\ 0 \text{ for } \xi > a \end{cases} \tag{1.35}$$

has spectrum function

$$F(\eta) = \frac{\sin(a\eta)}{\pi\eta}. \tag{1.36}$$

Here, of course, $F(\eta)$ has no singularity at $\eta = 0$, and the η path of integration runs undisturbed along the real axis.

Finally it is remarked that free use will be made of the concept of the delta function. This is written $\delta(\xi)$, and is as usual attributed with the formal definition

$$\delta(\xi) = 0 \quad \text{for} \quad \xi \neq 0, \quad \int_{-\infty}^{\infty} \delta(\xi)\, d\xi = 1. \tag{1.37}$$

On replacing $f(\xi)$ in (1.27) by $\delta(\xi)$ it appears that the spectrum function of $\delta(\xi)$ is simply $1/(2\pi)$, so that it has the formal representation

$$\delta(\xi) = \frac{1}{2\pi} \int_{-\infty}^{\infty} e^{i\xi\eta}\, d\eta. \tag{1.38}$$

A convenient way of thinking of the delta function in the present context is as the limit as $a \to 0$ of $1/(2a)$ times the rectangular pulse (1.35), since, loosely stated, this gives a rectangular pulse of unit area and zero width. The limit as $a \to 0$ of $1/(2a)$ times (1.36) of course recovers the spectrum $1/(2\pi)$.

The correctness of the relations (1.26) and (1.27) has been readily established for the rectangular pulse (1.35) and the associated spectrum (1.36) It is instructive to appreciate that this result can be made the basis of a heuristic demonstration of the validity of the relations for an effectively arbitrary function $f(\xi)$, in the following way. As just noted, (1.36) implies that $\delta(\xi)$ has spectrum $1/(2\pi)$. But $f(\xi)$ can be expressed as a superposition of delta functions; formally

$$f(\xi) = \int_{-\infty}^{\infty} f(\xi')\, \delta(\xi - \xi')\, d\xi'. \tag{1.39}$$

Hence the spectrum $F(\eta)$ of $f(\xi)$ is the corresponding superposition of the functions $\exp(-i\xi'\eta)/2\pi)$, these being the spectra of $\delta(\xi - \xi')$; that is,

$$F(\eta) = \frac{1}{2\pi} \int_{-\infty}^{\infty} f(\xi')\, e^{-i\xi'\eta}\, d\xi'.$$

CHAPTER II

PLANE WAVE REPRESENTATION

2.1. PLANE WAVES

2.1.1. Homogeneous Plane Waves in Vacuum

For a vacuum time-harmonic electromagnetic field of angular frequency ω that has space dependence only on the single rectangular cartesian coordinate x, Maxwell's equations for the complex representation of the field vectors in a region free of current are

$$H_x = 0, \quad \frac{\partial E_z}{\partial x} = i\omega\mu_0 H_y, \quad \frac{\partial H_y}{\partial x} = i\omega\varepsilon_0 E_z; \tag{2.1}$$

$$E_x = 0, \quad \frac{\partial H_z}{\partial x} = -i\omega\varepsilon_0 E_y, \quad \frac{\partial E_y}{\partial x} = -i\omega\mu_0 H_z. \tag{2.2}$$

These equations are, of course, the particular form assumed by (1.15) and (1.16) when **J** and the y derivatives are zero. Here, both **E** and **H** are transverse to Ox, and any field can be conceived as the superposition of two separate fields, in one of which E_z and H_y, in the other of which H_z and E_y, are the only non-zero field components. Furthermore, the latter field, governed by eqns. (2.2), in effect differs from the former, governed by eqns. (2.1), only in orientation with respect to the coordinate axes. The sole basic type of field can therefore be regarded as represented by that in which E_z and H_y, say, are the only non-zero components. Both E_z and H_y of course satisfy the one-dimensional wave equation

$$\frac{d^2\varphi}{dx^2} + k_0^2\varphi = 0, \tag{2.3}$$

where

$$k_0 = \omega\sqrt{\varepsilon_0\mu_0}, \tag{2.4}$$

and the radical, here and elsewhere, is taken to be positive.

Since two independent solutions of (2.3) are

$$e^{-ik_0x}, \quad e^{ik_0x}, \tag{2.5}$$

differing only in the sign of x, it is seen that the basic type of field is in effect

$$\mathbf{E} = (0, 0, 1)\, e^{-ik_0x}, \tag{2.6}$$

$$\mathbf{H} = Y_0\,(0, -1, 0)\, e^{-ik_0x}, \tag{2.7}$$

where

$$Y_0 = \sqrt{\varepsilon_0/\mu_0} \tag{2.8}$$

is the vacuum admittance, and for convenience the amplitude of the electric field has been normalized to unity.

Equations (2.6), (2.7) represent a linearly polarized, *homogeneous* plane wave travelling in the positive x-direction with speed

$$c = \omega/k_0 = 1/\sqrt{\varepsilon_0\mu_0} = 2{\cdot}998 \times 10^8 \text{ m sec}^{-1}.$$

The wavelength is

$$2\pi/k_0 = 2\pi c/\omega,$$

and the associated time-averaged power flux is

$$\text{Re}\, \tfrac{1}{2}\mathbf{E} \wedge \mathbf{H}^* = (\tfrac{1}{2}Y_0, 0, 0). \tag{2.9}$$

The term homogeneous is used in this context to signify that the equi-phase planes are also equi-amplitude planes; the condition that the field be independent of y and z does not, indeed, permit otherwise, even were some spatial variation of amplitude introduced by replacing the vacuum by a lossy medium. If, however, this condition is relaxed it is easy to establish the existence of solutions of Maxwell's equations that are still legitimately called plane waves in that the equi-phase surfaces are parallel planes, as are the equi-amplitude surfaces, but for which the two sets of planes face in different directions. Such plane waves are designated as *inhomogeneous*; they play an important part in the theory developed in this book, and are now investigated in some detail.

2.1.2. Inhomogeneous Plane Waves in Vacuum

The most general conditions are sought under which a vacuum, time-harmonic electromagnetic field in a current free region can have a space variation represented by

$$e^{-ik_0\mathbf{n}.\mathbf{r}}, \tag{2.10}$$

where k_0 is given by (2.4), $\mathbf{r} = (x, y, z)$ and $\mathbf{n}$ is some constant vector. As explained in § 1.2, a necessary condition is that (2.10) satisfy the time-harmonic wave equation

$$\nabla^2\varphi + k_0^2\varphi = 0, \tag{2.11}$$

and this will be the case if and only if

$$\mathbf{n}^2 = 1. \tag{2.12}$$

Moreover, since obtaining a solution of Maxwell's equations is synonymous with finding a divergence free vector whose components satisfy (2.11), the field can be written in the form

$$\mathbf{E} = \mathbf{A}e^{-ik_0\mathbf{n}\cdot\mathbf{r}}, \tag{2.13}$$

$$\mathbf{H} = Y_0\mathbf{n} \wedge \mathbf{A}e^{-ik_0\mathbf{n}\cdot\mathbf{r}}, \tag{2.14}$$

where $\mathbf{A}$ is some constant vector subject only to the relation

$$\mathbf{n} \cdot \mathbf{A} = 0. \tag{2.15}$$

When $\mathbf{n}$ is real its cartesian components are in effect, by virtue of (2.12), simply direction cosines; and (2.13), (2.14) then represent nothing other than a homogeneous plane wave, as encountered in the previous section, travelling in the direction $\mathbf{n}$, with $\mathbf{E}$, $\mathbf{H}$ and $\mathbf{n}$ mutually orthogonal.

There is, however, no requirement that the components of $\mathbf{n}$ be restricted to real values; for full generality they must be supposed complex. Write, accordingly,

$$\mathbf{n} = \mathbf{n}_r + i\mathbf{n}_i, \tag{2.16}$$

where $\mathbf{n}_r$ and $\mathbf{n}_i$ are real vectors, being respectively the real and imaginary parts of $\mathbf{n}$. The condition (2.12) then yields the two simultaneous conditions

$$\mathbf{n}_r^2 - \mathbf{n}_i^2 = 1, \tag{2.17}$$

$$\mathbf{n}_r \cdot \mathbf{n}_i = 0. \tag{2.18}$$

Relation (2.18) states that $\mathbf{n}_r$ and $\mathbf{n}_i$ are orthogonal, and with this knowledge the character of the plane wave (2.13), (2.14) is perhaps most readily exposed by taking two cartesian axes, Ox, Oy say, along $\mathbf{n}_r$, $\mathbf{n}_i$ respectively. Then full account is taken of (2.17) and (2.18) by writing

$$\mathbf{n}_r = (\cosh\beta, 0, 0), \ \mathbf{n}_i = (0, -\sinh\beta, 0), \tag{2.19}$$

where β is an arbitrary real parameter, positive, negative or zero.

With $\mathbf{A} = (A_x, A_y, A_z)$ the remaining condition (2.15) now reads

$$A_x \cosh\beta - iA_y \sinh\beta = 0, \tag{2.20}$$

which implies that

$$\mathbf{A} = (ia \sinh\beta,\ a\cosh\beta,\ b), \tag{2.21}$$

where a and b are arbitrary complex parameters.

Substitution from (2.19) and (2.21) into (2.13), (2.14) shows that the field is formed by linear combination of the two plane waves

$$\mathbf{E}, \mathbf{H} = \{(0, 0, 1),\ Y_0(-i\sinh\beta, -\cosh\beta, 0)\} \times e^{-k_0 y \sinh\beta}\, e^{-ik_0 x\cosh\beta}, \tag{2.22}$$

and

$$\mathbf{H}, \mathbf{E} = \{(0, 0, 1),\ Z_0\,(i\sinh\beta, \cosh\beta, 0)\} \times e^{-k_0 y \sinh\beta}\, e^{-ik_0 x\cosh\beta}. \tag{2.23}$$

The former results from putting $a = 0$, $b = 1$; the latter from putting $b = 0$, $a = Z_0$, where $Z_0 = 1/Y_0 = \sqrt{\mu_0/\varepsilon_0}$ is the vacuum impedance.

The plane waves (2.22) and (2.23) are inhomogeneous, with the equi-phase planes, $x =$ constant, at right angles to the equi-amplitude planes, $y =$ constant. The speed of phase propagation is $c/\cosh\beta$, less than the vacuum speed of light. There is a field component along the direction of phase propagation, Ox, and correspondingly a component of the real Poynting vector transverse to it. The time-averaged power flux, however, is directed strictly along Ox; for (2.22), for example,

$$\operatorname{Re}\tfrac{1}{2}\mathbf{E} \wedge \mathbf{H}^* = (\tfrac{1}{2}Y_0 \cosh\beta\, e^{-2k_0 y\sinh\beta}, 0, 0). \tag{2.24}$$

That the general plane wave, with phase propagation along Ox and amplitude variation along Oy, is expressible as a superposition of two independent plane waves, for which $\mathbf{E}$ and $\mathbf{H}$ respectively are everywhere parallel to Oz, is symptomatic of the fact, demonstrated in § 1.2, that any vacuum electromagnetic field that is independent of z can be expressed as a superposition of a field for which $\mathbf{E}$ is everywhere along Oz and a field for which $\mathbf{H}$ is everywhere along Oz. Evidently (2.22) and (2.23) are the respective E-polarized and H-polarized fields obtained by taking exponential plane wave solutions of the current free ($\mathbf{J} = 0$) forms of (1.15) and (1.16).

2.1.3. Plane Waves in an Isotropic Medium

As explained in § 1.2, it is supposed that the medium is specified by a certain permittivity ε and conductivity σ, which are basically functions of the angular frequency ω, though they may sometimes be regarded as substantially independent of ω; the permeability of the medium, on the other hand, is presumed not to differ significantly from the vacuum permeability μ_0.

When the medium is homogeneous, that is, when ε and σ are independent of position, the complex representation of any vacuum, time-harmonic electromagnetic field yields a field in the medium simply on replacing ε_0 by the complex permittivity

$$\varepsilon - i\sigma/\omega. \tag{2.25}$$

The most general type of plane wave in the medium is therefore represented by a superposition of the expressions derived from (2.22) and (2.23) by this substitution. Apart from the additional constant phase differences between the components of **E** and **H** consequent on the replacement of the ε_0 appearing in Y_0 and Z_0, the effect of the substitution is to modify the attenuation and phase propagation characteristics of the waves as defined by the exponential factors. Instead of $k_0 = \omega\sqrt{\varepsilon_0\mu_0}$ in these exponential factors there now appears $k_0\mu_c$, where μ_c is the complex refractive index, namely

$$\mu_c = \sqrt{\frac{\varepsilon}{\varepsilon_0} - \frac{i\sigma}{\omega\varepsilon_0}}. \tag{2.26}$$

On the presumption that σ, but not necessarily ε, is positive, (2.30) can be written

$$\mu_c = \mu - i\chi, \tag{2.27}$$

where μ and χ are positive, but otherwise can, at least in principle, assume any values; for to conform to the convention regarding k_0, stated after (2.4), that branch of the square root in (2.26) must be taken that is positive when σ is zero. The space dependence of the general plane wave in the medium is therefore specified by

$$e^{-k_0(\chi x \cosh\beta + \mu y \sinh\beta)}\, e^{-ik_0(\mu x \cosh\beta - \chi y \sinh\beta)}. \tag{2.28}$$

With the recollection that β is an arbitrary real parameter, it is seen at once that the direction of phase propagation ($\mu \cosh\beta$, $-\chi \sinh\beta$, 0) makes with the direction of maximum attenuation ($\chi \cosh\beta$, $\mu \sinh\beta$, 0) an angle γ that can take any value between

0 and $\frac{1}{2}\pi$. The expression for $\cos\gamma$ can be written

$$\cos\gamma = \left\{1 + \left[\left(\frac{\mu}{\chi} + \frac{\chi}{\mu}\right)\cosh\beta\sinh\beta\right]^2\right\}^{-1/2}. \tag{2.29}$$

If $\beta = 0$, then $\gamma = 0$ and (2.28) reverts to the familiar form

$$e^{-k_0\chi x}\, e^{-ik_0\mu x}, \tag{2.30}$$

which describes a homogeneous wave, analogous to that considered in § 2.1.1, travelling with speed of phase propagation c/μ in a conducting medium; in this case it is, of course, to be expected on physical grounds that for a medium with $\sigma > 0$ there will be attenuation in the direction of phase propagation. As β increases from zero and ultimately tends to infinity, so γ increases from zero and ultimately tends to $\frac{1}{2}\pi$.

2.1.4. Plane Waves in an Anisotropic Medium

If, at a fixed frequency, the linear dependence of the complex representation of **D** on that of **E** has perforce to be described by a tensor relation, say

$$\mathbf{D} = \varepsilon_0 \mathscr{K} \mathbf{E},$$

where $\mathscr{K}$ is the so-called dielectric tensor, then the properties of the plane wave solutions of Maxwell's equations are more diverse and a comprehensive investigation would be quite elaborate. For the most part this book is concerned only with vacuum fields, or fields in isotropic media, so the discussion in this section is confined to making and illustrating certain points without filling in all the details.

The permeability of the medium will be supposed not to differ from μ_0, so the complex representations of the time-harmonic field vectors satisfy eqns. (1.23), (1.24). If the space variation is specified by (2.10) these equations are

$$\mathbf{n} \wedge \mathbf{E} = Z_0 \mathbf{H}, \tag{2.31}$$

$$\mathbf{n} \wedge \mathbf{H} = -Y_0 \mathscr{K} \mathbf{E}, \tag{2.32}$$

and the substitution for **H** from (2.31) into (2.32) gives

$$\mathbf{n} \wedge (\mathbf{n} \wedge \mathbf{E}) = -\mathscr{K}\mathbf{E}, \tag{2.33}$$

that is

$$\mathbf{n}(\mathbf{n} \,.\, \mathbf{E}) + (\mathscr{K} - \mathbf{n}^2)\,\mathbf{E} = 0. \tag{2.34}$$

Equation (2.34) specifies three linear *homogeneous* equations for the three components of **E**, and thereby determines the condition on **n** under which there is a non-zero solution with space variation (2.10). The condition is, of course, the vanishing of the determinant formed by the coefficients of the components **E** in the three equations. To examine this explicitly, write

$$\mathbf{n} = (l, m, n), \tag{2.35}$$

where no confusion will arise from the use of the same letter for the z-component of the vector as for the vector itself, and introduce the tensor

$$\mathscr{N} = \begin{pmatrix} -m^2 - n^2 & lm & ln \\ ml & -l^2 - m^2 & mn \\ nl & nm & -l^2 - m^2 \end{pmatrix}. \tag{2.36}$$

Equation (2.34) now appears as

$$(\mathscr{N} + \mathscr{K})\,\mathbf{E} = 0, \tag{2.37}$$

and the condition is

$$\det(\mathscr{N} + \mathscr{K}) = 0. \tag{2.38}$$

To take the simplest possible case as an example, suppose that

$$\mathscr{K} = \begin{pmatrix} 1 & 0 & 0 \\ 0 & 1 & 0 \\ 0 & 0 & \varkappa \end{pmatrix}, \tag{2.39}$$

where $\varkappa$ is positive; then

$$\det(\mathscr{N} + \mathscr{K}) = \begin{vmatrix} 1 - m^2 - n^2 & lm & ln \\ ml & 1 - n^2 - l^2 & mn \\ nl & nm & \varkappa - l^2 - m^2 \end{vmatrix}. \tag{2.40}$$

The evaluation of the determinant (2.40) is facilitated by performing, sucessively, the following preliminary operations: subtract l/n times the third row from the first row; subtract m/n times the third row from the secound row; add l/n times the first column and m/n times the second column to the third column. The result is

$$\det(\mathscr{N} + \mathscr{K}) = \begin{vmatrix} 1 - l^2 - m^2 - n^2 & 0 & l(1 - \varkappa) \\ 0 & 1 - l^2 - m^2 - n^2 & m(1 - \varkappa) \\ l & m & \varkappa \end{vmatrix}$$

which is now easily seen to give

$$\det(\mathcal{N}+\mathcal{K}) = (1 - l^2 - m^2 - n^2)(\varkappa - l^2 - m^2 - \varkappa n^2). \tag{2.41}$$

The condition (2.38) is therefore equivalent to two alternative conditions; either

$$l^2 + m^2 + n^2 = 1, \tag{2.42}$$

or

$$l^2 + m^2 + \varkappa n^2 = \varkappa. \tag{2.43}$$

Condition (2.42), $\mathbf{n}^2 = 1$, is identical with (2.12), obtained for a vacuum plane wave. With regard to phase propagation and amplitude variation, therefore, the analysis of § 2.1.2 applies. It should, however, be remembered that in the present case the wave is also characterized by a rather particular polarization, as specified by the ratios of the components of **E** determined by (2.37) when (2.42) holds. In fact, it is easily deduced that, assuming $\varkappa$ to be neither zero nor unity, (2.37) and (2.42) imply

$$lE_x + mE_y = 0, \quad E_z = 0. \tag{2.44}$$

Since the expression of **H** in terms of **E**, namely (2.31), does not involve $\varkappa$, it follows that the entire structure of the plane wave is precisely that described in § 2.1.2, but that such plane waves can be propagated in the anisotropic medium if and only if there is no component of **E** in the direction, Oz, about which the anisotropy is symmetric.

Turn, now, to condition (2.43). If all the components of $\mathbf{n}$ are real the space variation (2.10) can be written

$$e^{-ik_0\mu\hat{\mathbf{n}}\cdot\mathbf{r}}, \tag{2.45}$$

where

$$\mu = \sqrt{l^2 + m^2 + n^2}, \tag{2.46}$$

and $\hat{\mathbf{n}} = \mathbf{n}/\mu$ is a real unit vector; this represents a homogeneous plane wave of uniform amplitude with equi-phase planes travelling in the direction $\hat{\mathbf{n}}$ with speed c/μ. The refractive index μ has a dependence on the direction of phase propagation determined by (2.43); and if θ is the angle that $\hat{\mathbf{n}}$ makes with the axis of symmetry, Oz, then $n/\mu = \cos\theta$, and the dependence is exhibited in the form

$$\mu^2 = \frac{\varkappa}{\sin^2\theta + \varkappa\cos^2\theta}. \tag{2.47}$$

To consider the general plane wave, however, $\mathbf{n}$ must be allowed to be complex. With $\mathbf{n} = \mathbf{n}_r + i\mathbf{n}_i$, as in § 2.1.2, and using suffices r and i likewise to denote the real and imaginary parts of the components (l, m, n), (2.43) yields the simultaneous relations

$$l_r l_i + m_r m_i + \varkappa n_r n_i = 0, \tag{2.48}$$

$$l_r^2 + m_r^2 + \varkappa n_r^2 - (l_i^2 + m_i^2 + \varkappa n_i^2) = \varkappa. \tag{2.49}$$

The relation (2.48) states that the direction in which the amplitude decays most rapidly, $-\mathbf{n}_i = (-l_i, -m_i, -n_i)$, must be perpendicular to the direction $(l_r, m_r, \varkappa n_r)$. In general, therefore, it is not perpendicular to the direction of phase propagation $\mathbf{n}_r = (l_r, m_r, n_r)$, but makes with this latter direction an angle which may be greater or less than $\frac{1}{2}\pi$, the extent of the possible excursion being dependent on the value of $\varkappa$. Moreover, this statement concerns only the ratios of the components of the various vectors and consequently needs no modification in the light of the additional relation (2.49), which may be regarded as simply placing a restriction on the amplitudes that can be associated with the ratios.

The reason why the direction in which the amplitude decays most rapidly is not perpendicular to the direction of phase propagation is because the latter no longer represents the direction of energy flow, an important general feature of propagation in an anisotropic medium. In a lossless medium the direction of amplitude decay would, in fact, be expected to be orthogonal to the time-averaged power flux $\mathrm{Re}\,\frac{1}{2}\mathbf{E} \wedge \mathbf{H}^*$. To conclude this section it is shown that for any medium specified by a Hermitian dielectric tensor this is indeed the case.

It is required to prove that

$$\mathbf{n}_i \, . \, (\mathbf{E} \wedge \mathbf{H}^*)_r = 0, \tag{2.50}$$

where the suffix r signifies the real part. Now

$$\mathbf{n} \, . \, (\mathbf{E} \wedge \mathbf{H}^*) = (\mathbf{n} \wedge \mathbf{E}) \, . \, \mathbf{H}^*,$$

and from (2.31) this is $Z_0 \mathbf{H} \, . \, \mathbf{H}^*$, which is real. Furthermore

$$\mathbf{n} \, . \, (\mathbf{E}^* \wedge \mathbf{H}) = -(\mathbf{n} \wedge \mathbf{H}) \, . \, \mathbf{E}^*,$$

and from (2.32) this is $Y_0(\mathscr{K}\mathbf{E}) \, . \, \mathbf{E}^*$, which again is real by virtue of $\mathscr{K}$ being Hermitian. Hence

$$\tfrac{1}{2}\mathbf{n} \, . \, (\mathbf{E} \wedge \mathbf{H}^* + \mathbf{E}^* \wedge \mathbf{H}) = \mathbf{n} \, . \, (\mathbf{E} \wedge \mathbf{H}^*)_r$$

is real, and the statement that its imaginary part is zero is precisely (2.50).

2.1.5. An Example

The concept of an unbounded plane wave is, of course, an idealization. Nevertheless, in theoretical work an unbounded homogeneous plane wave in a lossless medium is a convenient fiction which can be physically acceptable; scattering problems, for example, are often treated on the assumption that the field incident on the obstacle is such a plane wave. Unbounded inhomogeneous plane waves, on the other hand, represent a grosser violation of physical reality, since their exponential decay in one direction is matched by exponential growth in the opposite direction; it is unacceptable to allow the latter to develop without limit. It is therefore to be expected that an inhomogeneous plane wave can contribute to a field only throughout at most some half-space; and whereas in the representation of general fields as the superposition of plane waves, this feature is common to both homogeneous and inhomogeneous plane waves, it is also indicative of the situation which must be envisaged if a simple example involving a *single* inhomogeneous plane wave is sought.

To consider such an example, let the half-space $y > 0$ be vacuum and the half-space $y < 0$ be filled with a homogeneous lossless dielectric of permittivity $\varepsilon > \varepsilon_0$ and permeability μ_0. In the dielectric let there be incident on the interface $y = 0$ the homogeneous plane wave

$$\mathbf{E}^i, \mathbf{H}^i = \{(0, 0, 1),\ Y(\sin\alpha,\ -\cos\alpha, 0)\}\, e^{-ikr\cos(\theta-\alpha)}. \tag{2.51}$$

Equations (2.51) display the cartesian components of the field vectors; the xy-plane is the plane of incidence, the wave is polarized with the electric field parallel to the interface, and α is the (real) angle which the direction of propagation makes with the interface. The polar coordinates r, θ are given in terms of x, y by the relations

$$x = r\cos\theta, \quad y = r\sin\theta,$$

and

$$Y = \sqrt{\varepsilon/\mu_0}, \quad k = \omega\sqrt{\varepsilon\mu_0}.$$

To satisfy the boundary conditions at $y = 0$ and the conditions at $y = \pm\infty$ there must also be, in the dielectric, a reflected wave

$$\mathbf{E}^r, \mathbf{H}^r = \varrho\{(0, 0, 1),\ Y(-\sin\alpha,\ -\cos\alpha,\ 0)\}\, e^{-ikr\cos(\theta+\alpha)}, \tag{2.52}$$

and, in the vacuum, a transmitted wave

$$\mathbf{E}^t,\ \mathbf{H}^t = \tau\{(0, 0, 1),\ Y_0(\sin\gamma,\ -\cos\gamma,\ 0)\}\ e^{-ik_0 r\cos(\theta-\gamma)}, \tag{2.53}$$

where

$$k\cos\alpha = k_0\cos\gamma. \tag{2.54}$$

The reflection coefficient ϱ is determined directly by the continuity of impedance E_z/H_x at $y = 0$, which gives

$$\frac{1+\varrho}{Y\sin\alpha(1-\varrho)} = \frac{1}{Y_0\sin\gamma};$$

that is

$$\varrho = \frac{\mu\sin\alpha - \sin\gamma}{\mu\sin\alpha + \sin\gamma}, \tag{2.55}$$

where

$$\mu = \sqrt{\varepsilon/\varepsilon_0}.$$

And the transmission coefficient τ follows at once from the continuity of E_z, which gives

$$\tau = 1 + \varrho = \frac{2\mu\sin\alpha}{\mu\sin\alpha + \sin\gamma}. \tag{2.56}$$

Equation (2.54) can be written

$$\cos\gamma = \mu\cos\alpha. \tag{2.57}$$

If $\mu\cos\alpha \leqq 1$, then γ is real and the vacuum field is a homogeneous plane wave exhibiting refraction according to Snell's law. If, however, $\mu\cos\alpha > 1$, which obtains if α is less than the critical value $\cos^{-1}(1/\mu)$, then γ is pure imaginary and the vacuum field is an inhomogeneous plane wave; for this case the identification of $\mu\cos\alpha$ with $\cosh\beta$, where β is real, gives

$$\gamma = -i\beta, \tag{2.58}$$

and shows that (2.53) differs from (2.22) only by the constant amplitude-phase factor

$$\tau = \frac{2\mu\sin\alpha}{\mu\sin\alpha - i\sinh\beta}. \tag{2.59}$$

What has been described is, of course, nothing other than the theory of total internal reflection. Although there is a field in the vacuum, away from the interface it decays exponentially, and on average no power is transmitted across the interface. The wave in the vacuum is commonly called *evanescent*; it also illustrates the characteristics of so-called *surface waves*.

2.2. ANGULAR SPECTRUM OF PLANE WAVES

2.2.1. Plane Surface Currents

Since a field generated by a bounded current distribution is, in any direction, outgoing at infinity, it is to be expected that a single analytic representation as a superposition of plane waves will not, in general, be valid over more than a half-space; for a plane wave which travels outward in one direction travels inwards along the opposite direction, and one which decays in one direction grows in the opposite direction. It is convenient, therefore, to set up the theory of the representation by associating it with a surface density distribution of current flowing in a plane.

It is well known that, in crossing any surface current density **j**, the tangential components of **E** remain continuous, as does the tangential component of **H** parallel to **j**, but that the tangential component of **H** perpendicular to **j** jumps by the amount j. These statements, supplemented by proper allocation of sign to the discontinuity in **H**, may be expressed succinctly in the form

$$[\mathbf{n} \wedge \mathbf{E}] = 0, \tag{2.60}$$

$$[\mathbf{n} \wedge \mathbf{H}] = \mathbf{j}, \tag{2.61}$$

where **n** is the unit vector normal to the surface, and a quantity in square brackets denotes the increment which the quantity acquires in crossing the surface in the direction of **n**.

The result (2.60) follows from the integral form of (1.7), namely

$$\oint_l \mathbf{E} \,.\, d\mathbf{s} = -i\omega\mu_0 \int_S \mathbf{H} \,.\, d\mathbf{S}, \tag{2.62}$$

by taking the closed loop l, which is the rim of the surface S, to be a small narrow rectangle whose longer sides lie tangential to, and on either side of, the current carrying surface. Likewise (2.61) follows from the integral form of (1.8),

$$\oint_l \mathbf{H} \,.\, d\mathbf{s} = i\omega\varepsilon_0 \int_S \mathbf{E} \,.\, d\mathbf{S} + \int_S \mathbf{J} \,.\, d\mathbf{S}. \tag{2.63}$$

The behaviour of the normal component of **H** in crossing the surface current is determined by (2.60) and (1.7); it is continuous. The behaviour of the normal component of **E** is determined by (2.61) and (1.8); it jumps by the amount σ/ε_0, where σ is the surface charge density associated with **j** through the charge conservation relation. These results are commonly derived directly from (1.4) and (1.5) respectively, and are sometimes listed with (2.60) and (2.61) as through they were of equivalent status. For a proper appreciation of the basis of what is to follow it is important to recognize that (2.60) and (2.61) alone are fundamental conditions whose satisfaction must be ensured, in the sense that an electromagnetic field that is outgoing at infinity and satisfies those conditions *is* the unique field generated by the surface current, and cannot but exhibit the behaviour of the normal components of **E** and **H** just described.

It may also be remarked, in parenthesis, that the deduction of (2.60) and (2.61) from (2.62) and (2.63), respectively, depends on the assumption that **H** and **E** remain sufficiently well behaved in the vicinity of the current carrying surface for their fluxes across S to be neglected. Since a finite surface density of current itself represents an infinite volume density, this behaviour can hardly be taken for granted; it can, however, readily be established, for example, by treating the surface current as a volume density throughout a layer whose width is ultimately allowed to tend to zero.

Turning now to the specific case of surface currents flowing in a plane it only remains to draw attention to the nature of the symmetry of the field about the plane. If the plane is $z = 0$, then E_x, E_y and H_z have the same respective values at any point (x, y, z) as at the image point $(x, y, -z)$; whereas H_x, H_y and E_z have respective values at (x, y, z) which are the negative of those at $(x, y, -z)$. The jumps in H_x and H_y across $z = 0$ are therefore, in magnitude, twice the values that H_x and H_y respectively take just on one side of $z = 0$; and (2.61) states

$$j_x(x, y) = -2H_y(x, y, 0+), \tag{2.64}$$

$$j_y(x, y) = 2H_x(x, y, 0+). \tag{2.65}$$

That (2.60) is satisfied is implicit in the symmetry of E_x and E_y.

2.2.2. Angular Spectrum in Vacuum: Two-dimensional Case

It has been seen in § 2.1.2 that the most general type of time-harmonic plane wave in vacuum is essentially a two-dimensional field; (2.22) and (2.23) depend on x and y, but not on z. For this reason the main features of the representation of a vacuum field as a superposition of plane waves are incorporated in the analysis of a two-dimensional field. Since this case is relatively simple to expound, and is also adequate for a number of applications, it seems best to use it to introduce the theory, leaving the three-dimensional case to be set out subsequently as a straightforward generalization.

It is supposed, then, that the field is associated with a surface density of current flowing in the xz-plane, and that there is no dependence on z. Resolution into an E-polarized field and an H-polarized field is therefore possible, and each of these fields can be treated separately.

For the E-polarized field the current density in $y = 0$ has only a z-component, j_z say. Ultimately j_z is to be regarded as a virtually arbitrary function of x, but consider first the very simple case when it has some value, j_0, that is independent of x. Since in that case the field can only depend on y it must take the form of the homogeneous plane waves described in § 2.1.1, with spatial dependence $\exp(-ik_0y)$ or $\exp(ik_0y)$. Outgoing behaviour at infinity clearly demands that only the former can obtain in $y > 0$, and only the latter in $y < 0$. Taking account of the symmetry about $y = 0$ the field must therefore be

$$\mathbf{E} = -\tfrac{1}{2}Z_0 j_0(0, 0, 1)\, e^{\mp ik_0y}, \tag{2.66}$$

$$\mathbf{H} = \mp\tfrac{1}{2}j_0(1, 0, 0)\, e^{\mp ik_0y}, \tag{2.67}$$

where the upper/lower sign applies for $y \gtrless 0$, and the constant multiplying the exponential in (2.67) is fixed by the relation between the jump in $\mathbf{H}$ and the current density, which now reads (cf. (2.64), (2.65))

$$j_z = -2H_x(x, 0+). \tag{2.68}$$

The argument used in this very simple example is also immediately applicable when j_z has a harmonic variation with x. For evidently it is then only necessary to replace the plane waves (2.66), (2.67), which travel normal to $y = 0$, by plane waves travelling at the appropriate angle α to $y = 0$; that is

$$\mathbf{E} = -\tfrac{1}{2}Z_0 j_0(0, 0, 1)\, e^{-ik_0r\cos(\theta\mp\alpha)}, \tag{2.69}$$

$$\mathbf{H} = \tfrac{1}{2}j_0(\mp\sin\alpha, \cos\alpha, 0)\, e^{-ik_0r\cos(\theta\mp\alpha)}, \tag{2.70}$$

where the upper/lower sign applies for $y \gtrless 0$, and $x = r\cos\theta$, $y = r\sin\theta$. The sole component of the associated current density is, from (2.68),

$$j_z = j_0 \sin \alpha \, e^{-ik_0 x \cos\alpha}. \tag{2.71}$$

Now (2.71) represents a variation of current density with x that is harmonic, with wavelength $2\pi/(k_0 \cos \alpha)$. If, therefore, the wavelength of the current distribution exceeds the wavelength $2\pi/k_0$ of a homogeneous vacuum plane wave of angular frequency ω, then α is real and the field (2.69), (2.70) generated by the current distribution consists of a pair of homogeneous plane waves travelling away from the plane $y = 0$ into the respective half-spaces $y > 0$, $y < 0$. On average, energy is carried away from the plane $y = 0$, the time-averaged power flux across unit area of any plane $y =$ constant being

$$\tfrac{1}{8}Z_0|j_0|^2 \sin \alpha.$$

If, on the other hand, the wavelength of the current distribution is less than $2\pi/k_0$, then α is pure imaginary and the field consists of a pair of inhomogeneous plane waves. Specifically, if $\alpha = -i\beta$, (2.69), (2.70) are

$$\mathbf{E}, \mathbf{H} = -\tfrac{1}{2}Z_0 j_0\{(0, 0, 1),\ Y_0(\mp i \sinh \beta,\ -\cosh \beta, 0)\}\ e^{\mp k_0 y \sinh \beta} \times e^{-ik_0 x \cosh \beta},$$

which for $y > 0$ is precisely (2.22) multiplied by the constant amplitude-phase factor $-\tfrac{1}{2}Z_0 j_0$. The field is propagated in the x-direction with speed $c/\cosh \beta$ identical with that of the current density, and decays away from the plane $y = 0$ with decay coefficient $k_0 \sinh \beta$ which is larger the shorter the wavelength $2\pi/(k_0 \cosh \beta)$ of the current density. The time-averaged power flux is solely in the x-direction (see (2.24)), and on average, therefore, no energy is carried away from the plane $y = 0$.

The final step is now taken in supposing that j_z is essentially arbitrary. The only requirement is that it be expressible as a superposition of harmonic terms, that is, as a Fourier integral. Each harmonic term contributes to the field a pair of homogeneous or inhomogeneous plane waves; and since each plane wave is conveniently characterized, in the way just indicated, by a real or purely imaginary angle α, the complete field representation, obtained by superposition, may properly be called an *angular spectrum of plane waves*.

To put these ideas into mathematical form it is only necessary to note further that the expression of arbitrary j_z as a Fourier integral implies that $\cos\alpha$ must run the full range from ∞ to $-\infty$, so that α itself must cover the range defined by (*a*) $-i\beta$, as β goes through real values from ∞ to 0, (*b*) real values from 0 to π, (*c*) $\pi + i\beta$, as β goes through real values from 0 to ∞. The sign of the imaginary part of α in (*a*) and (*c*) does not affect the value of $\cos\alpha$, but is determined uniquely by the requirement that the inhomogeneous waves decay *away* from the plane $y = 0$; for example, it is seen in (2.22) that this condition is only satisfied if β is positive. The appropriate superposition of plane waves (2.69) and (2.70) can therefore be written in the form

$$\mathbf{E} = \int_C (0, 0, 1)\, P(\cos\alpha)\, e^{-ik_0 r\cos(\theta\mp\alpha)}\, d\alpha, \tag{2.72}$$

$$\mathbf{H} = Y_0 \int_C (\pm\sin\alpha, -\cos\alpha, 0)\, P(\cos\alpha)\, e^{-ik_0 r\cos(\theta\mp\alpha)}\, d\alpha, \tag{2.73}$$

with the upper/lower sign for $y \gtrless 0$, where the path of integration C in the complex α plane is that depicted in Fig. 2.1. The part of C lying along the real axis corresponds to homogeneous plane waves fanning out over the respective half-spaces, whereas the arms running parallel to the imaginary axis correspond to inhomogeneous plane waves decaying away from $y = 0$.

In (2.72), (2.73) $P(\cos\alpha)$ is the *spectrum function* which specifies in terms of amplitude and phase the "weight" attached to each plane wave of the spectrum. It is determined directly by the Fourier

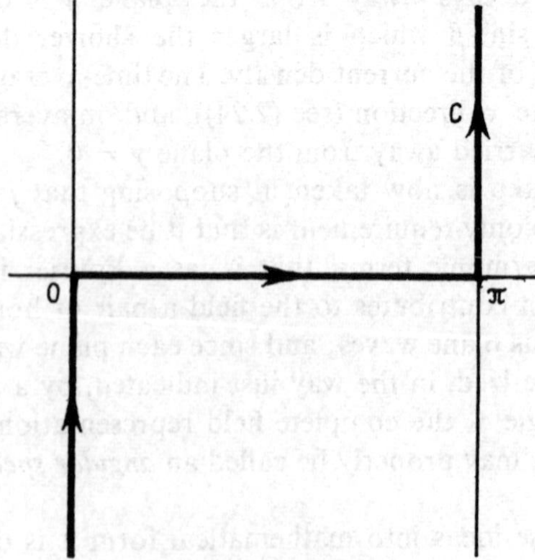

Fig. 2.1.

analysis of the current density, for the application of (2.73) to (2.68) gives

$$j_z(x) = -2Y_0 \int_C \sin\alpha \, P(\cos\alpha) \, e^{-ik_0 x \cos\alpha} \, d\alpha, \qquad (2.74)$$

which with the change of integration variable from α to

$$\lambda = \cos\alpha$$

is

$$j_z(x) = -2Y_0 \int_{-\infty}^{\infty} P(\lambda) \, e^{-ik_0 x\lambda} \, d\lambda. \qquad (2.75)$$

The formal expression for $P(\lambda)$ in terms of the current density is simply the inverse of (2.75), namely

$$P(\lambda) = -\frac{k_0 Z_0}{4\pi} \int_{-\infty}^{\infty} j_z(x) \, e^{ik_0 x\lambda} \, dx. \qquad (2.76)$$

To complete the representation of a two-dimensional field it remains to write down the form analogous to (2.72), (2.73) for an H-polarized field. This is taken as

$$\mathbf{H} = \pm \int_C (0, 0, 1) \, Q(\cos\alpha) \, e^{-ik_0 r \cos(\theta \mp \alpha)} \, d\alpha, \qquad (2.77)$$

$$\mathbf{E} = Z_0 \int_C (-\sin\alpha, \pm\cos\alpha, 0) \, Q(\cos\alpha) \, e^{-ik_0 r \cos(\theta \mp \alpha)} \, d\alpha, \qquad (2.78)$$

where the upper/lower sign applies for $y \gtrless 0$. The current density in the plane $y = 0$ now has only an x component, given by

$$j_x = 2H_z(x, 0+); \qquad (2.79)$$

thus

$$j_x = 2 \int_C Q(\cos\alpha) \, e^{-ik_0 x \cos\alpha} \, d\alpha, \qquad (2.80)$$

or, again taking $\lambda = \cos\alpha$,

$$j_x = 2 \int_{-\infty}^{\infty} \frac{Q(\lambda)}{\sqrt{1 - \lambda^2}} \, e^{-ik_0 x\lambda} \, d\lambda. \qquad (2.81)$$

The formal inverse is

$$\frac{Q(\lambda)}{\sqrt{1 - \lambda^2}} = \frac{k_0}{4\pi} \int_{-\infty}^{\infty} j_x(x) \, e^{ik_0 \lambda x} \, dx. \qquad (2.82)$$

The particular form in which (2.77), (2.78) are written brings out the mathematical parallelism between E-polarized fields and H-polarized fields that stems from the invariance of Maxwell's vacuum equations, in a current free region, under the tranformation $\mathbf{E} \to \mathbf{H}$, $\mathbf{H} \to -\mathbf{E}$, $\varepsilon_0 \leftrightarrow \mu_0$. A result of choosing this form is that it is the relation between the spectrum function $Q(\lambda)$ and the value of E_x at $y = 0$ that avoids the intrusion of the function $\sqrt{1 - \lambda^2}$ witnessed in (2.81), (2.82). Explicitly, from (2.78),

$$E_x(x, 0) = -Z_0 \int_{-\infty}^{\infty} Q(\lambda)\, e^{-ik_0 x\lambda}\, d\lambda, \tag{2.83}$$

and so

$$Q(\lambda) = -\frac{k_0 Y_0}{2\pi} \int_{-\infty}^{\infty} E_x(x, 0)\, e^{ik_0 x\lambda}\, dx. \tag{2.84}$$

The relation (2.84) gives the spectrum function as, in essence, the Fourier transform of the value in the "aperture plane" $y = 0$ of the x-component of the field; and (2.76) can, of course, by virtue of (2.68), be regarded in the same way. This suggests that, although the discussion has hitherto been developed specifically in terms of a current density as the source of the field, a slightly more flexible way of thinking is available, which may on occasion be used with advantage, by relating the plane wave resolution of a field in a half-space to the "aperture distribution" of certain tangential field components.

Another equivalent description is afforded by the introduction of the fictitious concept of a surface *magnetic* current density $\mathbf{j}_m$, say, for which the relations analogous to (2.60), (2.61) are $[\mathbf{n} \wedge \mathbf{H}] = 0$, $[\mathbf{n} \wedge \mathbf{E}] = -\mathbf{j}_m$. For example, the field (2.77), (2.78) in $y > 0$ can be regarded as produced by $j_{mz}(x)$ in $y = 0$; and then in (2.84) $E_x(x, 0)$ is replaced by $\frac{1}{2} j_{mz}(x)$ to yield the result analogous to (2.76).

Finally, a point is made concerning the options α and λ as integration variables. In addition to its more ready interpretation, the former has this advantage: that whereas the points $\lambda = \pm 1$ are, in general, branch points in the complex λ-plane, as evidenced, for example, in (2.81), the corresponding points in the complex α-plane are regular. The α-plane is exploited extensively in much of the subsequent work on two-dimensional problems. On the other hand, the direct application of known Fourier transform results in order to relate spectrum function and aperture distribution demands use of the λ-plane, and it is therefore necessary to know how the basic path of integration from $-\infty$ to ∞ avoids the branch points at $\lambda = \pm 1$. This is, in fact, made clear by considering the map of the complete λ-plane on the strip of the α-plane between $\operatorname{Re} \alpha = 0$ and $\operatorname{Re} \alpha = \pi$; the map is depicted in Fig. 2.2,

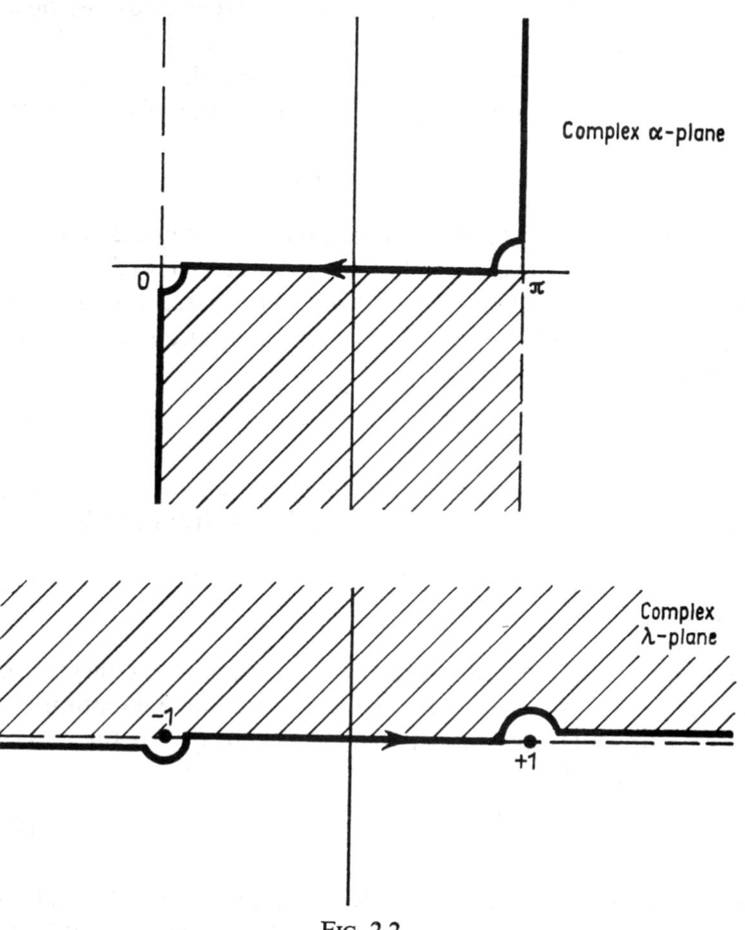

FIG. 2.2.

and exhibits the way in which the λ-path of integration must run in order to keep $\sqrt{1 - \lambda^2} = \sin \alpha$ negative pure imaginary when $|\lambda| > 1$. A warning should perhaps be added, to the effect that in the subsequent exploration of the complex α-plane beyond the confines $0 \leqq \operatorname{Re} \alpha \leqq \pi$ the continued use of the notation $P(\cos \alpha)$

must not be taken to imply that the function is necessarily either even in α or of period 2π.

2.2.3. Simple Examples: Line-sources

Some simple examples are now considered. Suppose, first, that the source of the field is an actual line current $I\cos(\omega t)$ flowing along the z-axis. The complex representation, leaving the time factor $\exp(i\omega t)$ understood, is simply I; and this is expressed as a surface current density (j_x, j_z) flowing in the plane $y = 0$ by taking

$$j_x = 0, \quad j_z = I\,\delta(x). \tag{2.85}$$

The delta function in (2.85) ensures that the current is zero everywhere but on the z-axis, and that the total current is I.

The field is, of course, E-polarized. The spectrum function is given by (2.76), and using (2.85) the integration is immediate, with the result

$$P(\lambda) = -\frac{k_0 Z_0 I}{4\pi}. \tag{2.86}$$

The plane waves of the spectrum therefore all have equal amplitudes, and are in phase at $r = 0$; specifically, from (2.72),

$$E_z = -\frac{k_0 Z_0 I}{4\pi}\int_C e^{-ik_0 r\cos(\theta\mp\alpha)}\,d\alpha. \tag{2.87}$$

In conjunction with this result it is interesting to consider solutions of the two-dimensional time-harmonic wave equation (1.17) that depend only on distance r from the z-axis. In this case the equation is simply

$$\frac{d^2\varphi}{dr^2} + \frac{1}{r}\frac{d\varphi}{dr} + k_0^2\varphi = 0; \tag{2.88}$$

that is, Bessel's equation of order zero and argument $k_0 r$. The solution of (2.88) that represents an outgoing wave at infinity is the Hankel function of the second kind, $H_0^{(2)}(k_0 r)$; for

$$H_0^{(2)}(k_0 r) \sim \sqrt{\frac{2}{\pi}}\, e^{\frac{1}{4}i\pi}\,\frac{e^{-ik_0 r}}{\sqrt{k_0 r}} \tag{2.89}$$

as $r \to \infty$. If the identification

$$E_z = E_0 H_0^{(2)}(k_0 r) \tag{2.90}$$

is made, it may be asked what is the source of the E-polarized field so defined. The source must be located at $r = 0$, that is along the z-axis, since everywhere else the homogeneous equation (2.88) is satisfied by E_z; furthermore, it must be a current flowing along the z-axis, since the field is E-polarized. It only remains to find the amount of current. This can be done by evaluating the integral (cf. (2.63))

$$\oint_l \mathbf{H} \,.\, d\mathbf{s}, \tag{2.91}$$

where the loop l is a circle in the plane $z = 0$, with centre the origin and vanishingly small radius ε. Evidently

$$\mathbf{H} \,.\, d\mathbf{s} = \varepsilon H_\theta \, d\theta;$$

and since

$$H_\theta = -\frac{iY_0}{k_0}\frac{\partial E_z}{\partial r} = -iY_0 E_0 H_0^{(2)\prime}(k_0 r), \tag{2.92}$$

(2.91) is

$$-2\pi i Y_0 E_0 \varepsilon H_0^{(2)\prime}(k_0 \varepsilon), \tag{2.93}$$

and it is required to find the limit of this expression as $\varepsilon \to 0$. Now the Hankel function behaves like the logarithm of its argument as its argument tends to zero, and the derivative of the Hankel function therefore behaves like the inverse of its argument. Specifically

$$H_0^{(2)\prime}(k_0\varepsilon) \sim -\frac{2i}{\pi k_0 \varepsilon} \quad \text{as} \quad \varepsilon \to 0, \tag{2.94}$$

and the substitution of (2.94) into (2.93) gives the total current

$$-4Y_0 E_0 / k_0. \tag{2.95}$$

It may be remarked that (2.92) can also be written

$$H_\theta = iY_0 E_0 H_1^{(2)}(k_0 r). \tag{2.96}$$

It has, then, been established that a current (2.95) flowing along the z-axis generates the E-polarized field defined by (2.90). Correspondingly, a current I gives

$$E_z = -\tfrac{1}{4} k_0 Z_0 I H_0^{(2)}(k_0 r). \tag{2.97}$$

This must be identical with (2.87), and hence

$$H_0^{(2)}(k_0 r) = \frac{1}{\pi}\int_C e^{-ik_0 r\cos(\theta \mp \alpha)}\, d\alpha. \tag{2.98}$$

In fact (2.98) is one of the standard integral representations of the Hankel function.

As a second example, closely connected with the first, take the problem of a slot in a conducting surface, idealized in the following way: suppose that a perfectly conducting sheet covers the entire plane $y = 0$ except for a slot of infinitesimal width cut along the z-axis, and let an actual voltage $V\cos(\omega t)$ be maintained across the slot. Then the resulting field is H-polarized, and can be determined by noting that the distribution of E_x in the plane $y = 0$ is given by the complex representation

$$E_x(x, 0) = V\delta(x). \tag{2.99}$$

For the substitution of (2.99) into (2.84) gives

$$Q(\lambda) = -\frac{k_0 Y_0 V}{2\pi}, \tag{2.100}$$

which in turn goes into (2.77) with the result

$$H_z = \mp\frac{k_0 Y_0 V}{2\pi}\int_C e^{-ik_0 r\cos(\theta\mp\alpha)}\,d\alpha = \mp\tfrac{1}{2}k_0 Y_0 V H_0^{(2)}(k_0 r). \tag{2.101}$$

In (2.101) the upper/lower sign applies, as usual, for $y \gtrless 0$, and the discontinuity in H_z in crossing $y = 0$ specifies the current density j_x that flows by virtue of the voltage maintained across the slot. Apart from constant factors, the solutions (2.87) and (2.101) are, of course, related by the transformation $\mathbf{E} \to \mathbf{H}$, $\mathbf{H} \to -\mathbf{E}$, $\varepsilon_0 \leftrightarrow \mu_0$.

The latter example illustrates that it may not always be most convenient to work directly in terms of the current distribution. If there *is* sought a current line-source that gives rise to an H-polarized field, it is necessary to postulate a current density

$$j_x = p\delta(x), \quad j_y = 0; \tag{2.102}$$

this is a line dipole of strength p (amp metre) per unit length (metre) of the z-axis. Then (2.82) gives

$$Q(\lambda) = \frac{k_0 p}{4\pi}\sqrt{1-\lambda^2}, \tag{2.103}$$

that is

$$Q(\cos\alpha) = \frac{k_0 p}{4\pi}\sin\alpha; \tag{2.104}$$

whence, from (2.77),

$$H_z = \pm \frac{k_0 p}{4\pi} \int_C \sin\alpha \, e^{-ik_0 r \cos(\theta \mp \alpha)} \, d\alpha. \tag{2.105}$$

The integral in (2.105) is $\pm i$ times the derivative with respect to $k_0 y$ of the integral in (2.98). Thus

$$H_z = \tfrac{1}{4} i k_0 p \sin\theta H_0^{(2)\prime}(k_0 r) = -\tfrac{1}{4} i k_0 p \sin\theta H_1^{(2)}(k_0 r). \tag{2.106}$$

2.2.4. Angular Spectrum in Vacuum: Three-dimensional Case

The principles involved should now be sufficiently clear for the plane wave spectrum representation of the vacuum field of a general time-harmonic distribution of surface current density flowing in a plane to be stated with no more than a few explanatory comments on the particular way it is written. The plane containing the current distribution is now taken to be $z = 0$, and the chosen form of the representation is

$$\mathbf{H} = \iint_{-\infty}^{\infty} \left[\pm P, \pm Q, \frac{lP + mQ}{\sqrt{1 - l^2 - m^2}} \right] \times e^{ik_0[lx + my \mp \sqrt{1 - l^2 - m^2} z]} \, dl \, dm, \tag{2.107}$$

$$\mathbf{E} = Z_0 \iint_{-\infty}^{\infty} \left[\frac{lmP + (1 - l^2)Q}{\sqrt{1 - l^2 - m^2}}, - \frac{(1 - m^2)P + mlQ}{\sqrt{1 - l^2 - m^2}}, \mp (mP - lQ) \right] e^{ik_0[lx + my \mp \sqrt{1 - l^2 - m^2} z]} \, dl \, dm, \tag{2.108}$$

with upper/lower sign for $z \gtrless 0$.

The following points are noted. First, that the directions of propagation of the plane waves of the spectrum are expressed in terms of (possibly complex) direction cosines $-l$, $-m$, $\pm\sqrt{1 - l^2 - m^2}$, rather than in terms of two (possibly complex) angles; this is because such angles of necessity enter in a somewhat unsymmetrical way, so that the paths of integration need rather careful examination; any gain from ease of physical interpretation or removal of branch points seems outweighed by this complication, at least in so far as the initial statement of the general theory is concerned.

The second point is that, for a general representation, there can be chosen just two independent spectrum functions associated with two of the six components of **E** and **H**, the spectrum functions of the remaining four components being then determined from Maxwell's equations; this is merely equivalent to the fact that any plane wave travelling in a given direction can be expressed uniquely as a superposition of two given independent plane waves travelling in that direction. Here the two independent spectrum functions $P(l, m)$, $Q(l, m)$ have been associated directly with H_x and H_y, the tangential components of **H**; that is, in effect, with the components of current density j_x, j_y. This choice accords with the point of view adopted in the introduction to the theory, and is the most convenient for present purposes. The spectrum function for H_z follows at once from div $\mathbf{H} = 0$, and those for the components of **E** are then obtained from curl $\mathbf{H} = i\omega\varepsilon_0\mathbf{E}$.

The explicit relations between P and Q and the current density come from (2.64) and (2.65); they are

$$j_x(x, y) = -2\iint_{-\infty}^{\infty} Q(l, m)\, e^{ik_0(lx+my)}\, dl\, dm, \tag{2.109}$$

$$j_y(x, y) = 2\iint_{-\infty}^{\infty} P(l, m)\, e^{ik_0(lx+my)}\, dl\, dm. \tag{2.110}$$

These double Fourier transforms can be thought of as repeated single Fourier transforms. The inverses are

$$P(l, m) = \frac{k_0^2}{8\pi^2}\iint_{-\infty}^{\infty} j_y(x, y)\, e^{-ik_0(lx+my)}\, dx\, dy, \tag{2.111}$$

$$Q(l, m) = -\frac{k_0^2}{8\pi^2}\iint_{-\infty}^{\infty} j_x(x, y)\, e^{-ik_0(lx+my)}\, dx\, dy. \tag{2.112}$$

The plane waves corresponding to those parts of the paths of integration for l and m on which $l^2 + m^2 > 1$ are, of course, inhomogeneous, and correct behaviour at infinity demands that $\sqrt{1 - l^2 - m^2}$ is then negative pure imaginary.

Finally it may be remarked that if the current plane $z = 0$ is situated not in a vacuum but in an isotropic medium as specified in § 2.1.3, then the required modification of (2.107) and (2.108) is formally quite trivial. It is only necessary to replace each radical $\sqrt{1 - l^2 - m^2}$ by $\sqrt{\mu_c^2 - l^2 - m^2}$; to replace the factors $1 - l^2$ and $1 - m^2$ multiplying Q and P in the spectra of E_x and E_y by $\mu_c^2 - l^2$ and $\mu_c^2 - m^2$, respectively; and to replace Z_0 in the expression for **E** by Z_0/μ_c^2. The representation of **H** in the plane $z = 0$ is unaltered, and equations (2.111), (2.112) for the spectrum functions in terms of the current density continue to hold.

2.2.5. Simple Example: Dipole Source

A simple example of the three-dimensional spectrum is afforded by the fundamental point-source of the electromagnetic field, namely the electric dipole. In general this would be defined by the volume current density

$$\mathbf{J} = \mathbf{p}\,\delta(x - x_0)\,\delta(y - y_0)\,\delta(z - z_0), \qquad (2.113)$$

where the dipole is located at (x_0, y_0, z_0), is directed along **p**, and has "strength" p (amp metre). In the present context it is treated as a surface current density (j_x, j_y) in the plane $z = 0$; and if, for convenience, it is located at the origin and directed along the x-axis, then

$$j_x = p\delta(x)\,\delta(y), \quad j_y = 0. \qquad (2.114)$$

Substitution from (2.114) into (2.111) and (2.112) gives the corresponding spectrum functions

$$P = 0, \quad Q = -\frac{pk_0^2}{8\pi^2}. \qquad (2.115)$$

The field vectors follow from (2.107) and (2.108). In particular

$$\mathbf{H} = -\frac{pk_0^2}{8\pi^2}\int\!\!\int_{-\infty}^{\infty}\left[0, \pm 1, \frac{m}{\sqrt{1 - l^2 - m^2}}\right]$$

$$\times\, e^{ik_0[lx + my \mp \sqrt{1 - l^2 - m^2}z]}\,dl\,dm. \qquad (2.116)$$

To relate the double integral in (2.116) to the conventional expression for the dipole field it is perhaps simplest to note that (2.116) can be written

$$H_x = 0, \quad H_y = ik_0 Y_0\frac{\partial M}{\partial z}, \quad H_z = -ik_0 Y_0\frac{\partial M}{\partial y}, \qquad (2.117)$$

where

$$M = -\frac{pZ_0}{8\pi^2}\iint_{-\infty}^{\infty} \frac{1}{\sqrt{1-l^2-m^2}}\, e^{ik_0[lx+my\mp\sqrt{1-l^2-m^2}z]}\, dl\, dm. \tag{2.118}$$

Now (2.117) is $\mathbf{H} = ik_0 Y_0 \operatorname{curl} \mathbf{M}$, with $\mathbf{M} = (M, 0, 0)$. $\mathbf{M}$ is therefore the electric Hertz vector, and the standard result is

$$M = -\frac{ipZ_0}{4\pi}\frac{e^{-ik_0 r}}{k_0 r}. \tag{2.119}$$

The identification of (2.118) and (2.119) gives

$$2\pi i \frac{e^{-ik_0 r}}{k_0 r} = \iint_{-\infty}^{\infty} \frac{1}{\sqrt{1-l^2-m^2}}\, e^{ik_0[lx+my\mp\sqrt{1-l^2-m^2}z]}\, dl\, dm. \tag{2.120}$$

The formula (2.120) is important, whether regarded as the representation of a spherical wave by a superposition of plane waves, or as the evaluation of a double integral. It is equivalent to another standard result which, since it is required later, is conveniently established here.

Make the change of integration variables

$$l = \tau \cos\psi, \quad m = \tau \sin\psi,$$

and at the same time write

$$x = \varrho \cos\chi, \quad y = \varrho \sin\chi.$$

Then the right-hand side of (2.120) is, for $z > 0$,

$$\int_0^{\infty}\int_{2\pi \text{ range}} e^{ik_0\varrho\tau\cos(\psi-\chi)}\, d\psi\, \frac{\tau e^{-ik_0 z\sqrt{1-\tau^2}}}{\sqrt{1-\tau^2}}\, d\tau.$$

Now the ψ integral is $2\pi J_0(k_0\varrho\tau)$ (cf. (2.98)), so (2.120) states that

$$\int_0^{\infty} J_0(k_0\varrho\tau)\frac{\tau e^{-ik_0 z\sqrt{1-\tau^2}}}{\sqrt{1-\tau^2}}\, d\tau = \frac{ie^{-ik_0\sqrt{\varrho^2+z^2}}}{k_0\sqrt{\varrho^2+z^2}}, \tag{2.121}$$

which is the result in question.

2.2.6. Angular Spectrum in an Anisotropic Medium

In this section brief consideration is given to the new features that arise when the current density in $z = 0$ is radiating into a homogeneous anisotropic medium specified by a dielectric tensor as in § 2.1.4. In the present description, for the sake of simplicity, it is assumed that the anisotropy is such that the nature of the symmetry of the field about $z = 0$ is the same as that for the vacuum field, described in § 2.2.1.

It is convenient to retain as close contact as possible with (2.107) by introducing plane waves specified by

$$\mathbf{H} = \left(\pm P, \pm Q, \frac{lP + mQ}{n}\right) e^{ik_0(lx+my\mp z)}; \tag{2.122}$$

this is the most general form compatible with the particular condition div $\mathbf{H} = 0$, which condition, from (1.23), still stands. Now, however, the expression for n in terms of l and m is given by (2.38). This is a fourth degree equation for n, and with the present assumptions about symmetry is in fact a quadratic for n^2 (see, for example, (2.41)) yielding two independent forms, n_1 and n_2, say, for insertion into (2.122). For each form of n the ratios of the three components of $\mathbf{E}$ are given in terms of l and m by (2.37). Likewise for $\mathbf{H}$. Thus for $n = n_1$ the ratio P/Q is specified as a certain function of l and m, and again for $n = n_2$ as some other function of l and m.

It appears, then, that the representation analogous to (2.107), (2.108) can be written

$$\mathbf{H} = \mathbf{H}_1 + \mathbf{H}_2, \tag{2.123}$$

where

$$\mathbf{H}_i = \iint_{-\infty}^{\infty} \left(\pm P_i, \pm Q_i, \frac{lP_i + mQ_i}{n_i}\right) e^{ik_0(lx+my\mp n_i z)}\, dl\, dm, \quad (i = 1, 2); \tag{2.124}$$

the surface current density is

$$(j_x, j_y) = 2 \iint_{-\infty}^{\infty} (-Q_1 - Q_2, P_1 + P_2)\, e^{ik_0(lx+my)}\, dl\, dm, \tag{2.125}$$

with the inverse relations

$$(P_1 + P_2, Q_1 + Q_2) = \frac{k_0^2}{8\pi^2} \iint_{-\infty}^{\infty} (j_y, -j_x)\, e^{-ik_0(lx+my)}\, dx\, dy, \tag{2.126}$$

and P_1, P_2, Q_1, Q_2 are determined from (2.126) and the prescribed ratios P_1/Q_1, P_2/Q_2.

One important point to remember is that, to satisfy the radiation condition, for each homogeneous plane wave of the spectrum that solution for n_i must be chosen which ensures that the time-averaged power flux Re $\frac{1}{2}\mathbf{E} \wedge \mathbf{H}^*$ is directed away from the plane $z = 0$. For an anisotropic medium this does not necessarily mean that the corresponding direction of phase propagation is away from the plane $z = 0$; it is possible for it to be towards the plane, in which case n_i in (2.124) is negative. The nature of the relationship between the directions of phase propagation and power fiux are liable to be quite complicated, but would have to be understood in any particular problem under investigation.

In conclusion it may be remarked that if the medium, or its orientation with respect to the current distribution, are not such that the fields above and below $z = 0$ are related in the way implied by (2.124), then the dependence of n on l and m is specified by four (in general, independent) functions n_i^+, n_i^- ($i = 1, 2$), say, two of which are associated with the field in $z > 0$, the other two with that in $z < 0$. Correspondingly, there are eight spectrum functions P_i^+, Q_i^+, P_i^-, Q_i^-. The four ratios P_i^+/Q_i^+, P_i^-/Q_i^- are prescribed, and the remaining four equations for the determination of the spectrum functions comprise the two that identify j_x, j_y with the discontinuities in H_y, H_x across $z = 0$, and the two that state that E_x and E_y are continuous across $z = 0$.

CHAPTER III

SUPPLEMENTARY THEORY

3.1. RADIATED POWER

3.1.1. The Two-dimensional Case

In the present chapter the account of the basic theory given in Chapter 2 is supplemented by certain important developments of a general character. The main concern is to investigate the nature and limits of validity of the approximation leading to the radiation field. First, however, expressions for the time-averaged radiated power are obtained by appealing directly to the plane wave representation.

Consider the two-dimensional E-polarized field (2.72), (2.73). The time-averaged power flux across any plane y-constant (>0, say) is, per unit length in the z-direction,

$$\mathrm{Re}\,\tfrac{1}{2}\int_{-\infty}^{\infty} E_z H_x^*\,dx; \tag{3.1}$$

and the substitution into (3.1) of the spectrum representation gives

$$\mathrm{Re}\,\tfrac{1}{2}Y_0 \int_{-\infty}^{\infty}\int_C\int_C \sin^*\alpha'\,P(\cos\alpha)\,P^*(\cos\alpha')\,e^{-ik_0x(\cos\alpha-\cos\alpha')} \times e^{-ik_0y(\sin\alpha-\sin^*\alpha')}\,d\alpha\,d\alpha'\,dx, \tag{3.2}$$

where it is recalled that, along the contour C, $\cos\alpha$ is real and therefore equal to its conjugate complex, but that $\sin\alpha$ can be either real or pure imaginary. If the x integration is carried out, using (1.38), (3.2) becomes

$$\mathrm{Re}\,\frac{\pi Y_0}{k_0}\int_C\int_C \delta(\cos\alpha - \cos\alpha')\sin^*\alpha'\,P(\cos\alpha)\,P^*(\cos\alpha') \times e^{-ik_0y(\sin\alpha-\sin^*\alpha')}\,d\alpha\,d\alpha'. \tag{3.3}$$

The delta function in the integrand means that contributions are made to the double integral only when α and α' are equal. Moreover if the common value of α and α' is real, then $\sin^* \alpha' = \sin \alpha$, and the exponential factor in the integrand is unity; whereas if the common value is complex with real part zero or π, then $\sin^* \alpha' = -\sin \alpha$ is pure imaginary, and the entire integrand (apart from the delta function) with $\alpha = \alpha'$ is therefore also pure imaginary. Thus (3.3) is simply

$$\frac{\pi Y_0}{k_0} \int_0^\pi |P(\cos \alpha)|^2 \, d\alpha. \tag{3.4}$$

The result may be interpreted by saying that the time-averaged power radiated in any direction α is $(\pi Y_0/k_0)\,|P(\cos \alpha)|^2$ per unit angular spread, per unit length in the z-direction. That the expression (3.4) is indepedent of y could have been anticipated, because every unbounded plane y = constant subtends at $x = y = 0$ the full angular spread π. The expression (3.4) gives the power radiated into the half-space $y > 0$, and is therefore half the total power radiated. With $y = 0$, the initial expression (3.1) is the same as the form

$$\operatorname{Re} \tfrac{1}{2} \int_{-\infty}^{\infty} E_z j_z^* \, dx$$

for the total power radiated.

An entirely similar calculation applies to the H-polarized field (2.77), (2.78). In this case the power radiated into each half-space, per unit length in the z-direction, is

$$\frac{\pi Z_0}{k_0} \int_0^\pi |Q(\cos \alpha)|^2 \, d\alpha. \tag{3.5}$$

It is instructive to apply (3.4) and (3.5) to the simple examples of § 2.2.3. For the case in which current I flows along the z-axis, the substitution of (2.86) into twice (3.4) gives radiated power

$$\tfrac{1}{8} Z_0 k_0 I^2, \tag{3.6}$$

per unit length in the z-direction. This can be checked against the more conventional method in which the radiation field is used. From (2.97) and (2.89),

$$E_z \sim -\frac{1}{4} \sqrt{\frac{2}{\pi}}\, Z_0 k_0 I \, e^{\frac{1}{4} i\pi} \frac{e^{-ik_0 r}}{\sqrt{k_0 r}}, \tag{3.7}$$

and this, of course, is also the value in the radiation field of $-Z_0H_\theta$. The corresponding expression for the outward radial power flux density is

$$\text{Re} - \tfrac{1}{2}E_zH_\theta^* \sim \frac{Z_0k_0I^2}{16\pi r}. \tag{3.8}$$

The total radiated power follows from integration over unit length of the surface of a circular cylinder, axis the z-axis, radius r; and since this merely corresponds to multiplication of (3.8) by $2\pi r$, (3.6) is at once recovered.

Again, for the voltage slot (2.99), the substitution (2.100) into twice (3.5) gives radiated power

$$\tfrac{1}{2}Y_0k_0V^2; \tag{3.9}$$

whereas for the line dipole (2.102) the spectrum function is (2.104), and the radiated power is

$$\frac{1}{16}Z_0k_0p^2. \tag{3.10}$$

3.1.2. The Three-dimensional Case

The power radiated by an arbitrary current density flowing in the plane $z = 0$ is now considered. The time-averaged power flux across any plane $z =$ constant (>0, say) is

$$\text{Re}\,\tfrac{1}{2}\iint_{-\infty}^{\infty}(E_xH_y^* - E_yH_x^*)\,dx\,dy. \tag{3.11}$$

If the representations (2.107), (2.108) are fed into (3.11), the x- and y-integrations can be carried out and give delta functions; the result, which is analogous to (3.3), can be written

$$\text{Re}\,\frac{2\pi^2Z_0}{k_0^2}\iiiint_{-\infty}^{\infty}\delta(l-l')\,\delta(m-m')$$
$$\times\frac{lm(PQ^* + QP^*) + (1-m^2)PP^* + (1-l^2)QQ^*}{\sqrt{1-l^2-m^2}}$$
$$\times e^{-ik_0z[\sqrt{1-l^2-m^2}-\sqrt{1-l'^2-m'^2}]}\,dl\,dm\,dl'\,dm', \tag{3.12}$$

where it is understood that P and Q each have arguments (l, m), and that P^* and Q^* each have arguments (l', m').

The delta functions in (3.12) mean that contributions are made to the quadruple integral only when both $l = l'$ and $m = m'$. Moreover, if the common value of l and l', and the common value of m and m', are such that $l^2 + m^2 < 1$, then $\sqrt{1 - l^2 - m^2}$ is real and the exponential factor in the integrand of (3.12) is unity; whereas, if $l^2 + m^2 > 1$, then $\sqrt{1 - l^2 - m^2}$ is pure imaginary and the entire integrand (apart from the delta functions) with $l = l'$, $m = m'$ is pure imaginary. The time-averaged power crossing z = constant is therefore

$$\frac{2\pi^2 Z_0}{k_0^2} \iint\limits_{l^2+m^2<1} \frac{lm(PQ^* + QP^*) + (1 - m^2)PP^* + (1 - l^2)QQ^*}{\sqrt{1 - l^2 - m^2}} \times dl\, dm. \tag{3.13}$$

Since the integration in (3.13) is over the region $l^2 + m^2 < 1$ the interpretation of l and m as direction cosines is straightforward. With

$$l = -\sin\alpha\cos\beta, \quad m = -\sin\alpha\sin\beta, \tag{3.14}$$

(3.13) is transformed to

$$\frac{2\pi^2 Z_0}{k_0^2} \int\limits_{\beta=0}^{2\pi} \int\limits_{\alpha=0}^{\frac{1}{2}\pi} [\sin^2\alpha\sin\beta\cos\beta(PQ^* + QP^*) + (1 - \sin^2\alpha\sin^2\beta)PP^* + (1 - \sin^2\alpha\cos^2\beta)QQ^*]\sin\alpha\, d\alpha\, d\beta. \tag{3.15}$$

In (3.14) α and β are, of course, the polar angles, referred to the z-axis, that specify the direction of propagation of the plane waves of the spectrum. Since $\sin\alpha\, d\alpha\, d\beta$ is an element of solid angle the implication of (3.15) is that the power radiated, per unit solid angle in any direction α, β, is $2\pi^2 Z_0/k_0^2$ times the part of the integrand in square brackets.

These findings can be illustrated by the electric dipole, for which the spectrum functions are given in (2.115). Taking twice (3.15), the radiated power appears as

$$\frac{Z_0 k_0^2 p^2}{16\pi^2} \int\limits_0^{2\pi} \int\limits_0^{\frac{1}{2}\pi} (1 - \sin^2\alpha\cos^2\beta)\sin\alpha\, d\alpha\, d\beta;$$

and with the double integral easily found to be $4\pi/3$, this is

$$\frac{Z_0 k_0^2}{12\pi} p^2, \tag{3.16}$$

the familiar result.

3.2. THE RADIATION FIELD

3.2.1. Heuristic Approach: Stationary Phase

The current source of any actual time-harmonic field can be presumed to be contained in some region which is finite in all its dimensions. At sufficiently great distances from such a source it is possible to isolate a dominant part of the field, namely that part whose amplitude falls off as the inverse of the distance; this dominant contribution is known as the *radiation field* (or *far field*) of the source. It is now proposed to substantiate this statement (and its two-dimensional counterpart), to derive expressions for the radiation field from the plane wave representation, and to examine rather carefully the conditions under which the expressions do indeed give an adequate approximation to the complete field.

In this subsection a preliminary, heuristic argument is presented, based on what is commonly called the method of *stationary phase*. Consider, first, the representation (2.72) of the two-dimensional E-polarized field. For $y > 0$,

$$E_z = \int_C P(\cos\alpha)\, e^{-ik_0 r\cos(\theta-\alpha)}\, d\alpha. \tag{3.17}$$

Now at points where $k_0 r \gg 1$, the amplitudes of the inhomogeneous waves of the spectrum are very small, and such waves may be neglected. Moreover, the contributions of the homogeneous waves largely annul each other by destructive interference, since with $k_0 r \gg 1$ the phase of the waves in general varies rapidly with α, in the sense that a phase change of π is achieved by only a small change in α. Exceptionally, however, those waves for which α is close to θ interfere constructively, since the variation of phase with α vanishes at $\alpha = \theta$. Thus it can be argued that only that part of the path of integration C in the vicinity of $\alpha = \theta$ contributes significantly to (3.17), and consequently an approximation to (3.17) is

$$P(\cos\theta) \int_C e^{-ik_0 r\cos(\theta-\alpha)}\, d\alpha. \tag{3.18}$$

As found in (2.98), the integral in (3.18) is $\pi H_0^{(2)}(k_0 r)$, which in turn is naturally replaced by the form (2.89) valid for $k_0 r \gg 1$. Finally, then, the expression for E_z in the radiation field is

$$E_z \sim \sqrt{2\pi}\, e^{\frac{1}{4}i\pi} P(\cos\theta) \frac{e^{-ik_0 r}}{\sqrt{k_0 r}}, \tag{3.19}$$

representing an outgoing cylindrical wave with a "polar diagram" specified by $P(\cos\theta)$.

As a check it may be observed that (3.19) leads to

$$\text{Re} - \tfrac{1}{2} E_z H_\theta^* = \frac{\pi Y_0}{k_0 r} |P(\cos\theta)|^2$$

for the power flux density, which gives the power radiated into the half-space $y > 0$ as

$$\frac{\pi Y_0}{k_0} \int_0^\pi |P(\cos\theta)|^2 \, d\theta,$$

in agreement with (3.4).

The same heuristic argument can be applied to the three-dimensional representation (2.107). It is easy to verify the expected result, that at the point (x, y, z) with spherical polar coordinates (r, θ, φ) the phase of the homogeneous waves is stationary with respect to variations in both l and m when

$$l = -\frac{x}{r} = -\sin\theta\cos\varphi, \quad m = -\frac{y}{r} = -\sin\theta\sin\varphi. \tag{3.20}$$

The approximation to (2.107) analogous to (3.18) may therefore be written, for $z > 0$,

$$\mathbf{H} = [P\cos\theta, Q\cos\theta, -\sin\theta(P\cos\varphi + Q\sin\varphi)]$$
$$\times \iint_{-\infty}^{\infty} \frac{e^{ik_0[lx+my-\sqrt{1-l^2-m^2}z]}}{\sqrt{1-l^2-m^2}}\, dl\, dm. \tag{3.21}$$

In this expression P and Q are evaluated at (3.20); moreover, in view of (2.120), it is obviously convenient to preserve the factor $1/\sqrt{1-l^2-m^2}$ in the integrand, and it may be noted that $\sqrt{1-l^2-m^2}$ evaluated at (3.20) is simply $\cos\theta$. Finally, then, using (2.120) and transforming to the spherical polar components of **H**, the expressions for these in the radiation field are seen to be

$$(H_r, H_\theta, H_\varphi) \sim 2\pi i[0, P\cos\varphi + Q\sin\varphi, -\cos\theta\,(P\sin\varphi - Q\cos\varphi)] \times \frac{e^{-ik_0 r}}{k_0 r}, \tag{3.22}$$

It is easy to check that the corresponding time-averaged power flux density

$$\tfrac{1}{2}Z_0(H_\theta H_\theta^* + H_\varphi H_\varphi^*) \tag{3.23}$$

yields (3.15) when integrated over the appropriate hemisphere.

For plane aperture distributions that are spatially bounded, in the sense that they can be completely enclosed by some circle of finite radius, the associated spectrum functions, being integrals such as (2.111) over finite ranges, are free of singularities. The rate of change of each spectrum function therefore remains finite over the entire spectrum, and the method of stationary phase leading to (3.18) and (3.21) can undoubtedly be validated by taking $k_0 r$ so large that the dominance of the explicit exponential functions in controlling the behaviour of the integrands of (2.72) and (2.107), in the way noted in the remarks following (3.17), is complete. This confirms the well-known result, stated at the beginning of this subsection, that for a given spatially bounded source there is some finite distance beyond which the expressions for the radiation field are good approximations to the complete field.

In practice, however, it may well be that the points at which it is desired to evaluate the field are not sufficiently distant to be in the radiation field, even though $k_0 r \gg 1$. In the context of the plane wave spectrum representation this is broadly interpreted in the following way: the more extensive is the aperture distribution the more "peaked" does the spectrum function become in some part of the spectrum, with the result that, for a given value of r, the behaviour of the spectrum function may be comparably important to that of the exponential in determining the outcome of the plane wave superposition expressed by the integral representation.

Furthermore, it is sometimes theoretically convenient to consider an infinitely extended aperture distribution, in which case the spectrum function can have a singularity, with the result that, in certain directions, the concept of the radiation field is inapplicable no matter how distant the point of observation.

Even when the concept of the radiation field is not valid it may be possible to obtain an approximation to the plane wave representation of the field based solely on the condition $k_0r \gg 1$. With this in view the method of *steepest descents* is next introduced, being a procedure in some respects akin to that of stationary phase, but leading more readily to a rigorous analysis.

3.2.2. Rigorous Approach: Steepest Descents

The method of stationary phase accepts the integral (3.17) as it stands, and appeals to the concept of phase interference to justify the assertion that when $k_0r \gg 1$ only that part of C in the vicinity of $\alpha = \theta$ contributes significantly to the integral. In contrast, the method of steepest descents proceeds by first distorting the original path of integration C into a new path everywhere along which $ik_0r[1 - \cos(\theta - \alpha)]$ is real. Before examining the effect of this it is worth pausing to consider what distortions of C are legitimate, leaving it understood that proper allowance is made for the presence of any singularities that $P(\cos\alpha)$ may have.

If the real and imaginary parts of α are displayed explicitly thus

$$\alpha = \alpha_r + i\alpha_i, \tag{3.24}$$

then

$$-ik_0r\cos(\alpha - \theta) = -ik_0r\cos(\alpha_r - \theta)\cosh\alpha_i - k_0r\sin(\alpha_r - \theta) \times \sinh\alpha_i. \tag{3.25}$$

It is therefore necessary for the convergence of the integral that $\sin(\theta - \alpha_r)\sinh\alpha_i$ be negative when $\alpha_i \to \pm\infty$; and consequently the extremes of any path of integration obtained by distorting C must lie in the shaded sectors of Fig. 3.1, which are specified by

$$\theta < \alpha_r < \theta + \pi \quad \text{when} \quad \alpha_i > 0,$$
$$-\pi + \theta < \alpha_r < \theta \quad \text{when} \quad \alpha_i < 0.$$

To come now to the specific possibility that a path can be found such that everywhere on it $ik_0r[1 - \cos(\theta - \alpha)]$ is real, it need only be noted that the requirement, evident from (3.25), is simply

$$\cos(\alpha_r - \theta)\cosh\alpha_i = 1. \tag{3.26}$$

By comparing the graphs of $\cos x$ and $\operatorname{sech} x$ it is readily seen that the path specified by (3.26) passes through $\alpha = \theta$ at an angle $\frac{1}{4}\pi$ to the axes, and has $\alpha_r = \theta + \frac{1}{2}\pi$ and $\alpha_r = \theta - \frac{1}{2}\pi$ as asymptotes;

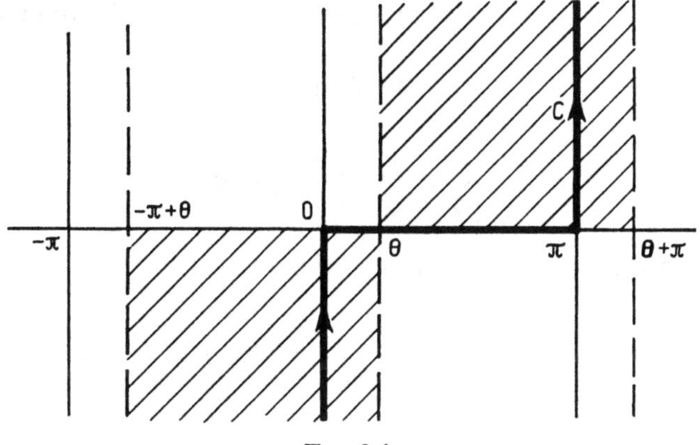

FIG. 3.1.

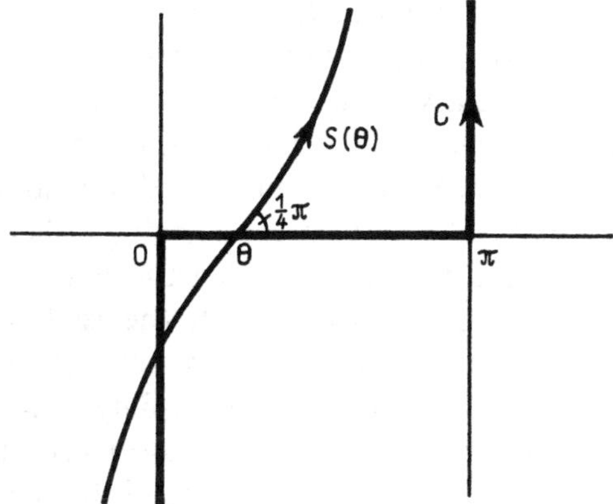

FIG. 3.2.

it is shown diagrammatically in Fig. 3.2, and is henceforth designated as $S(\theta)$. Moreover, (3.26) is equivalent to

$$\sin(\alpha_r - \theta) = \tanh \alpha_i, \tag{3.27}$$

and when (3.26) and (3.27) are obeyed (3.25) states

$$-ik_0 r \cos(\alpha - \theta) = -ik_0 r - k_0 r \sinh \alpha_i \tanh \alpha_i, \tag{3.28}$$

Hence on $S(\theta)$ the real part of (3.28) has its maximum value, zero, at $\alpha = \theta$, where $\alpha_i = 0$, and decreases monotonically to $-\infty$ away from $\alpha = \theta$ on either side. It is therefore possible, after distorting the path to $S(\theta)$, to change the variable of integration from α to τ, where

$$-ik_0 r \cos(\alpha - \theta) = -ik_0 r - k_0 r \tau^2, \tag{3.29}$$

and τ runs through real values from $-\infty$ to ∞. Evidently (3.29) is equivalent to

$$\tau = \sqrt{2}\, e^{-\frac{1}{4}i\pi} \sin [\tfrac{1}{2}(\alpha - \theta)], \tag{3.30}$$

so that

$$d\tau = \frac{1}{\sqrt{2}} e^{-\frac{1}{4}i\pi} \sqrt{1 - \tfrac{1}{2}i\tau^2}\, d\alpha, \tag{3.31}$$

and the explicit transformation of the integral is

$$\int\limits_{S(\theta)} P(\cos\alpha)\, e^{-ik_0 r \cos(\theta-\alpha)}\, d\alpha = \sqrt{2}\, e^{\frac{1}{4}i\pi}\, e^{-ik_0 r} \int\limits_{-\infty}^{\infty} \frac{P(\cos\alpha)}{\sqrt{1 - \frac{1}{2}i\tau^2}}\, e^{-k_0 r \tau^2}\, d\tau, \tag{3.32}$$

where on the right-hand side $P(\cos\alpha)$ is understood to be interpreted as the corresponding function of τ.

In the discussion of the previous section $\alpha = \theta$ was important as the point of stationary phase; in the present discussion it is important as a *saddle-point*. That is to say, the real part of $-ik_0 r \times \cos(\theta - \alpha)$, and hence the modulus of the exponential of this quantity, is stationary at $\alpha = \theta$, without being either a true maximum or a true minimum. $S(\theta)$ is called the path of steepest descents, implying that path along which the real part of $-ik_0 r \times \cos(\theta - \alpha)$ decreases most rapidly as α proceeds away from the saddle-point; the quantity increases most rapidly along a path orthogonal to $S(\theta)$ at $\alpha = \theta$, and remains constant (zero) along the real axis. The merit of the steepest descents path is basically that, when $k_0 r \gg 1$, only that part of the path in the vicinity of the saddle-point contributes significantly to the integral. In this respect its importance is akin to that of the stationary phase path. It has, however, an added advantage in securing the transformation (3.32), since this form leads directly to the complete asymptotic expansion in descending powers of $k_0 r$, as is now shown.

An asymptotic expansion of any function $f(z)$ of a complex variable z is defined as follows: if

$$f(z) = \sum_{m=0}^{n} \frac{a_m}{z^m} + R_n(z), \tag{3.33}$$

where, for all n, $z^n R_n(z) \to 0$ as $|z| \to \infty$ for arg z within a given interval, then

$$a_0 + \frac{a_1}{z} + \frac{a_2}{z^2} + \cdots \tag{3.34}$$

is called an asymptotic expansion of $f(z)$ for the given range of arg z.

There is no implication that (3.34) is convergent, and indeed asymptotic expansions are often divergent for all values of the argument. The point is that for sufficiently large values of $|z|$ the terms of the series decrease at least initially, and that if the series is truncated before the smallest term the error is of the order of magnitude of the first discarded term.

The result to be proved can now be stated thus: if $F(\tau)$ is any function for which

$$\int_0^\infty F(\tau)\, e^{-\varkappa\tau^2}\, d\tau \tag{3.35}$$

converges for sufficiently large positive values of the parameter $\varkappa$, then the asymptotic expansion of (3.35) in descending powers of $\sqrt{\varkappa}$ is given by replacing $F(\tau)$ by its Taylor series in ascending powers of τ and integrating term by term. Explicitly, if

$$F(\tau) = \sum_{s=0}^{\infty} a_s \tau^s \quad \text{for} \quad |\tau| < \tau_0, \tag{3.36}$$

τ_0 being the radius of convergence of the Taylor series, then, since

$$\int_0^\infty \tau^s\, e^{-\varkappa\tau^2}\, d\tau = \frac{1}{2}\left(\frac{s-1}{2}\right)!\, \varkappa^{-\frac{1}{2}(s+1)}, \tag{3.37}$$

it is claimed that

$$\frac{1}{2\sqrt{\varkappa}} \sum_{s=0} \left(\frac{s-1}{2}\right)! \frac{a_s}{\varkappa^{\frac{1}{2}s}} \tag{3.38}$$

is the asymptotic expansion of (3.35).

The proof is as follows. It is presumed that the integral (3.35) converges for sufficiently large positive values of $\varkappa$, and that there is a number d such that $\exp(-d\tau^2)\, F(\tau)$ is bounded for all positive values of τ. Then, for $\tau > 0$,

$$\left| F(\tau) - \sum_{s=0}^{n-1} a_s \tau^s \right| < K\tau^n\, e^{d\tau^2},$$

where K is some fixed positive number. Hence

$$\int_0^\infty \left[F(\tau) - \sum_{s=0}^{n-1} a_s \tau^s\right] e^{-\varkappa\tau^2}\, d\tau$$

$$< K \int_0^\infty \tau^n\, e^{-(\varkappa-d)\tau^2}\, d\tau$$

$$= \frac{1}{2} K \left(\frac{n-1}{2}\right)! \, (\varkappa - d)^{-\frac{1}{2}(n+1)},$$

so that (3.38) is indeed an asymptotic expansion of (3.35) in accordance with the definition (3.33).

In the application of this result to (3.32) it need only be noted that since the range of integration in (3.32) runs from $-\infty$ to ∞ the integrals of the odd terms in the development of $P(\cos\alpha) \times (1 - \frac{1}{2}i\tau^2)^{-\frac{1}{2}}$ as a power series in τ vanish; the powers of $k_0 r$ in the final asymptotic expansion are therefore $-\frac{1}{2}, -\frac{3}{2}, -\frac{5}{2}, \ldots$ The first term evidently gives the approximation

$$\int_{S(\theta)} P(\cos\alpha)\, e^{-ik_0 r\cos(\theta-\alpha)}\, d\alpha \sim \sqrt{2\pi}\, e^{\frac{1}{4}i\pi}\, P(\cos\theta) \frac{e^{-ik_0 r}}{\sqrt{k_0 r}}, \tag{3.39}$$

confirming (3.19), and incidentally proving the asymptotic form of the Hankel function (2.89) which was originally quoted as a standard result. Indeed, a good illustration of the method of steepest descents is afforded by the Hankel representation (2.98). Since the integrand has no singularities the path C can be replaced by $S(\theta)$ without more ado, and the representation can then be transformed as in (3.32) with $P(\cos\alpha)$ replaced by $1/\pi$; the full asymptotic expansion is therefore obtained by developing $(1-\frac{1}{2}i\tau^2)^{-\frac{1}{2}}$ as a power series in τ and carrying out the term by term integration.

It has, then, been rigorously established that (3.39) is the first term of the asymptotic expansion of (3.32) in inverse powers of $\sqrt{k_0 r}$. It is now necessary, as previously explained, to inquire more closely into the question of how large $k_0 r$ must be in order that (3.39) be an adequate approximation to (3.32). The answer is to be found in the nature of the development of $P(\cos\alpha)$ as a power series in τ, as may be appreciated by looking again at the asymptotic expansion (3.38) of (3.35). Although eventually the factor $[\frac{1}{2}(s-1)]!$ may well become dominant, it can be said that the condition for initial successive terms to decrease markedly is $\sqrt{\varkappa} \gg a_s/a_{s+1}$, and this is the required criterion.

The point is well illustrated by the common situation in which the radius of convergence τ_0 of the power series (3.36) is finite; then broadly speaking the dependence of the coefficients a_s on s goes like τ_0^s, and the condition is $\varkappa \gg 1/\tau_0^2$. For example, in the asymptotic expansion of the Hankel function just mentioned the power series in τ is that of $(1 - \frac{1}{2}i\tau^2)^{-\frac{1}{2}}$, which has radius of convergence $\sqrt{2}$; the condition is therefore $k_0 r \gg \frac{1}{2}$. It is clear that for any case in which τ_0 is finite the asymptotic expansion cannot be convergent.

It turns out that in a number of problems the spectrum function $P(\cos \alpha)$ contains a simple pole, and that this particular singularity plays a major role in determining the nature of the field. The next section is devoted to an examination of this important special case.

3.3. ANGULAR SPECTRUM WITH SIMPLE POLE

3.3.1. The Complex Fresnel Integral

It will be seen shortly that the plane wave representation in which the spectrum function has a simple pole can be expressed in terms of a function which it is here convenient to write in the particular form

$$F(a) = e^{ia^2} \int_a^\infty e^{-i\tau^2}\, d\tau. \tag{3.40}$$

Some of the properties of this function are now briefly enunciated.

In (3.40) the argument a is conceived as taking arbitrary complex values, and the factor $\exp(ia^2)$ has been included in the definition of the function in order to secure boundedness as $|a| \to \infty$ when arg a lies in the range $-\pi$ to $\frac{1}{2}\pi$, which covers the cases of physical interest considered subsequently.

It is also convenient to have an explicit notation for the related function

$$F_0(a) = e^{ia^2} \int_0^a e^{-i\tau^2}\, d\tau. \tag{3.41}$$

Since

$$\int_{-\infty}^\infty e^{-i\tau^2}\, d\tau = 2\int_0^\infty e^{-i\tau^2}\, d\tau = \sqrt{\pi}\, e^{-\frac{1}{4}i\pi} \tag{3.42}$$

it is evident that

$$F(a) + F_0(a) = \tfrac{1}{2}\sqrt{\pi}\, e^{-\frac{1}{4}i\pi}\, e^{ia^2}, \tag{3.43}$$

$$F(a) + F(-a) = \sqrt{\pi}\, e^{-\frac{1}{4}i\pi}\, e^{ia^2}. \tag{3.44}$$

When a is real, $F_0(a)$ is expressible in terms of the *Fresnel* integrals, first introduced in the early optical theory of diffraction. When arg $a = -\frac{1}{4}\pi$, (3.40) becomes

$$F(|a|\, e^{-\frac{1}{4}i\pi}) = e^{-\frac{1}{4}i\pi}\, e^{|a|^2} \int_{|a|}^{\infty} e^{-\tau^2}\, d\tau,$$

expressible in terms of the *error* integral. And when arg $a = \frac{1}{4}\pi$, (3.41) is

$$F_0(|a|\, e^{\frac{1}{4}i\pi}) = e^{\frac{1}{4}i\pi}\, e^{-|a|^2} \int_0^{|a|} e^{\tau^2}\, d\tau, \tag{3.45}$$

from which (3.43) gives

$$F(|a|\, e^{\frac{1}{4}i\pi}) = -e^{\frac{1}{4}i\pi}\, e^{-|a|^2} \left[\int_0^{|a|} e^{\tau^2}\, d\tau + \tfrac{1}{2}i\sqrt{\pi}\right]. \tag{3.46}$$

Another form of the complex Fresnel integral is provided by

$$I = b \int_{-\infty}^{\infty} \frac{e^{-\varkappa\tau^2}}{\tau^2 + ib^2}\, d\tau, \tag{3.47}$$

where $\varkappa$ is a real positive parameter which could be absorbed in b, but whose retention serves a useful purpose, particularly in the proof now to be given of the relation between (3.47) and (3.40). Since

$$\frac{d}{d\varkappa}(I\, e^{-ib^2\varkappa}) = -b \int_{-\infty}^{\infty} e^{-(\tau^2 + ib^2)\varkappa}\, d\tau$$

$$= -\sqrt{\pi}\, b\, \frac{e^{-i\varkappa b^2}}{\sqrt{\varkappa}},$$

it follows that

$$I = \sqrt{\pi}\, b\, e^{ib^2\varkappa} \int_{\varkappa}^{\infty} \frac{e^{-ib^2\varkappa'}}{\sqrt{\varkappa'}}\, d\varkappa', \tag{3.48}$$

the upper limit being determined by the fact that $I \to 0$ as $\varkappa \to \infty$; and if momentarily b is taken to be real and positive the transformation $\varkappa' = \tau^2/b^2$ shows that (3.48) leads to

$$I = 2\sqrt{\pi}\, e^{ib^2\varkappa} \int\limits_{b\sqrt{\varkappa}}^{\infty} e^{-i\tau^2}\, d\tau = 2\sqrt{\pi}\, F(b\sqrt{\varkappa}). \tag{3.49}$$

Thus far (3.49) has been proved for b real and positive. Since (3.47) is an odd function of b, the corresponding result for b real and negative is

$$I = -2\sqrt{\pi}\, F(-b\sqrt{\varkappa}). \tag{3.50}$$

Furthermore, regarded as a function of b, (3.47) is obviously regular everywhere in the complex b plane except where arg b is either $\frac{1}{4}\pi$ or $-\frac{3}{4}\pi$. By analytic continuation, therefore,

$$b \int\limits_{-\infty}^{\infty} \frac{e^{-\varkappa\tau^2}}{\tau^2 + ib^2}\, d\tau = \pm\, 2\sqrt{\pi}\, F(\pm b\sqrt{\varkappa}), \tag{3.51}$$

where $\varkappa$ is real and positive, and where the upper sign holds for $-\frac{3}{4}\pi < \arg b < \frac{1}{4}\pi$, the lower for $\frac{1}{4}\pi < \arg b < 5\pi/4$.

Series representations of $F(a)$ or $F_0(a)$ are available. A convergent expansion in ascending powers of a can be obtained by repeated integration by parts, starting from

$$\int\limits_0^a e^{-i\tau^2}\, d\tau = [\tau\, e^{-i\tau^2}]_0^a + 2i \int\limits_0^a \tau^2\, e^{-i\tau^2}\, d\tau.$$

This process gives

$$F_0(a) = a \sum_{n=0}^{\infty} \frac{(2ia^2)^n}{1.\,3.\,5\,\ldots\,(2n+1)}. \tag{3.52}$$

An alternative approach is simply to use the exponential series under the integral sign in (3.41), and then to integrate term by term. This gives

$$F_0(a) = e^{ia^2}\, a \sum_{n=0}^{\infty} \frac{(-ia^2)^n}{n!(2n+1)}. \tag{3.53}$$

The asymptotic expansion of $F(a)$ in descending powers of a can also be obtained by repeated integration by parts. In this case the starting point is

$$\int\limits_a^{\infty} e^{-i\tau^2}\, d\tau = \left[-\frac{e^{-i\tau^2}}{2i\tau}\right]_a^{\infty} - \frac{1}{2i} \int\limits_a^{\infty} \frac{e^{-i\tau^2}}{\tau^2}\, d\tau,$$

and the process gives

$$F(a) = \frac{1}{2ia}\left[1 + \sum_{m=1}^{n} \frac{1.\,3.\,5 \ldots (2m-1)}{(-2ia^2)^m}\right] + R_n, \tag{3.54}$$

where

$$R_n = \frac{1.\,3.\,5 \ldots (2n+1)}{(-2i)^{n+1}} e^{ia^2} \int_a^\infty \frac{e^{-i\tau^2}}{\tau^{2n+2}}\, d\tau. \tag{3.55}$$

When a is real and positive

$$|R_n| \leqq \frac{1.\,3.\,5 \ldots (2n+1)}{2^{n+1}} \int_a^\infty \frac{d\tau}{\tau^{2n+2}}$$

$$= \frac{1.\,3.\,5 \ldots (2n-1)}{2^{n+1}} \frac{1}{a^{2n+1}},$$

so that from (3.54)

$$F(a) \sim \frac{1}{2ia}\left[1 + \frac{1}{(-2ia^2)} + \frac{1.\,3}{(-2ia^2)^2} + \cdots\right] \tag{3.56}$$

is indeed an asymptotic expansion. When a is real and negative the corresponding result comes from (3.44), namely

$$F(a) \sim \sqrt{\pi}\, e^{-\frac{1}{4}i\pi} e^{ia^2} + \frac{1}{2ia}\left[1 + \frac{1}{(-2ia^2)} + \frac{1.\,3}{(-2ia^2)^2} + \cdots\right]. \tag{3.57}$$

It is not difficult to generalize the analysis just given to show directly that (3.56) in fact holds for $-\frac{3}{4}\pi < \arg a < \frac{1}{4}\pi$, and (3.57) for $\frac{1}{4}\pi < \arg a < 5\pi/4$. It is also instructive to note that an alternative derivation is available by the application of the method of steepest descents to the left-hand side of (3.51), in precisely the manner discussed in § 3.2.2. It is only necessary to develop $(\tau^2 + ib^2)^{-1}$ as an ascending power series in τ, and then to integrate term by term; the influence of the finite radius of convergence of the power series in τ is apparent, and the resulting asymptotic expansion is divergent, as is indeed evident on inspection. It only remains to note that the change from the form (3.56) to the form (3.57) takes place on the *Stokes lines*, here $\arg a = \frac{1}{4}\pi$ and $\arg a = -\frac{3}{4}\pi$, where the difference between the two forms assumes its smallest value for a given value of $|a|$.

The early terms in (3.52) or (3.53) give useful approximations when $|a| \ll 1$, those in (3.56) and (3.57) when $|a| \gg 1$. Even when the accuracy required is not great there remains an appreciable gap to be bridged. Some ingenious analytic approximations have been proposed in the past for special applications, but more recently the need has been adequately met by the tabulation for complex values of the argument of several functions effectively equivalent to $F(a)$.

3.3.2. Reduction to Fresnel Integral

Consider the particular plane wave representation

$$\int_{S(\theta)} \sec\left(\frac{\alpha - \alpha_0}{2}\right) e^{-ik_0 r \cos(\theta - \alpha)}\, d\alpha, \tag{3.58}$$

in which the only singularities of the spectrum function in the complex α-plane are simple poles at $\alpha = \alpha_0 \pm (2n-1)\pi$, $n = 1, 2, 3, \ldots$, and the path of integration is presumed already distorted from C to $S(\theta)$, with proper allowance understood for any pole captured in the process.

By changing the variable of integration from α to $\alpha - \theta$ (3.58) appears, as

$$\int_{S(0)} \sec\left(\frac{\alpha - \alpha_0 + \theta}{2}\right) e^{-ik_0 r \cos\alpha}\, d\alpha; \tag{3.59}$$

or, by now reversing the sign of α, as

$$\int_{S(0)} \sec\left(\frac{\alpha + \alpha_0 - \theta}{2}\right) e^{-ik_0 r \cos\alpha}\, d\alpha. \tag{3.60}$$

The addition of (3.59) and (3.60), and division by two, then puts (3.58) in the form

$$2\cos\left(\frac{\alpha_0 - \theta}{2}\right) \int_{S(0)} \frac{\cos(\frac{1}{2}\alpha)}{\cos\alpha + \cos(\alpha_0 - \theta)} e^{-ik_0 r \cos\alpha}\, d\alpha. \tag{3.61}$$

Now, following (3.30), make the change of variable

$$\tau = \sqrt{2}\, e^{-\frac{1}{4}i\pi} \sin(\tfrac{1}{2}\alpha). \tag{3.62}$$

Then (3.61) becomes

$$-2e^{-\frac{1}{4}i\pi} e^{-ik_0 r}\, b \int_{-\infty}^{\infty} \frac{e^{-k_0 r \tau^2}}{\tau^2 + ib^2}\, d\tau, \tag{3.63}$$

where

$$b = \sqrt{2}\cos\left(\frac{\theta - \alpha_0}{2}\right). \tag{3.64}$$

But (3.63) can be expressed in terms of the complex Fresnel integral through (3.51). Finally, therefore,

$$\int_{S(\theta)} \sec\left(\frac{\alpha - \alpha_0}{2}\right) e^{-ik_0 r\cos(\theta-\alpha)}\,d\alpha$$

$$= \mp 4\sqrt{\pi}\, e^{-\frac{1}{4}i\pi}\, e^{-ik_0 r} F\left[\pm\sqrt{2k_0 r}\cos\left(\frac{\theta-\alpha_0}{2}\right)\right], \tag{3.65}$$

with the upper sign for $\theta - \alpha_0$ between $S(-\pi)$ and $S(\pi)$, and the lower sign otherwise, remembering, of course, that the expression has period 4π in $\theta - \alpha_0$.

The result (3.65) is exact, and is important as a "canonical" formula that describes precisely what happens when, for a given value of $k_0 r$ much greater than unity, the asymptotic approximation (3.39) nevertheless fails because of the close approach of one of the poles to the saddle-point. The approximation (3.39) corresponds to using the first term of (3.56) in (3.65), and the condition for its validity is

$$\left|\sqrt{2k_0 r}\cos\left(\frac{\theta-\alpha_0}{2}\right)\right| \gg 1. \tag{3.66}$$

The way in which the canonical formula can be used to give uniform asymptotic approximations to more general plane wave representations that are characterized by spectra containing simple poles is now described.

3.3.3. Steepest Descents with Saddle-point Near a Pole

Consider once again

$$\int_{S(\theta)} P(\cos\alpha)\, e^{-ik_0 r\cos(\theta-\alpha)}\,d\alpha, \tag{3.67}$$

and suppose now that $P(\cos\alpha)$ has a simple pole at $\alpha_0 - \pi$ that may approach arbitrarily close to the saddle-point θ. Suppose also that no other singularities need special consideration; in particular, that there are none near $\alpha_0 - \pi$. Then for $k_0 r \gg 1$ a uniform asymptotic approximation, that is, an approximation which remains valid for all values of $\theta - \alpha_0$, can be obtained by writing

$$P(\cos\alpha) = P_1(\cos\alpha) + p\sec\left(\frac{\alpha - \alpha_0}{2}\right), \tag{3.68}$$

where p is independent of α, and is so chosen that the function P_1 has no pole at $\alpha_0 - \pi$. In fact

$$p = \lim_{\alpha \to \alpha_0 - \pi} \cos\left(\frac{\alpha - \alpha_0}{2}\right) P(\cos\alpha), \tag{3.69}$$

and the resolution (3.68) simply splits off the pole in a way suggested by the canonical result (3.65). By hypothesis, P_1 contains no singularities requiring special treatment, and the uniform asymptotic approximation to (3.67) is therefore

$$\sqrt{2\pi}\, e^{\frac{1}{4}i\pi} P_1(\cos\theta) \frac{e^{-ik_0 r}}{\sqrt{k_0 r}}$$
$$\mp 4\sqrt{\pi}\, p\, e^{-\frac{1}{4}i\pi} e^{-ik_0 r} F\left[\pm\sqrt{2k_0 r}\cos\left(\frac{\theta - \alpha_0}{2}\right)\right], \tag{3.70}$$

with upper or lower sign as in (3.65).

A slightly simpler version of this result can be obtained by a very similar if less immediately convincing argument. Instead of being split off as in (3.68) the pole is factored out by writing

$$P(\cos\alpha) = \sec\left(\frac{\alpha - \alpha_0}{2}\right) P_2(\cos\alpha). \tag{3.71}$$

It is then argued that since $P_2(\cos\alpha)$ has no singularities in the vicinity of the saddle-point it may be removed from under the integral sign with α equated to θ. This step can indeed be justified quite straightforwardly by going over to the τ integral as in (3.32), and developing as a Taylor series in ascending powers of τ what remains of the relevant part of the integrand after the pole has been factored out. The resulting approximation to (3.67) is

$$\mp 4\sqrt{\pi}\, e^{-\frac{1}{4}i\pi} P_2(\cos\theta)\, e^{-ik_0 r} F\left[\pm\sqrt{2k_0 r}\cos\left(\frac{\theta - \alpha_0}{2}\right)\right]. \tag{3.72}$$

For $k_0 r \gg 1$, expressions (3.70) and (3.72) are in close agreement, in that when the modulus of the argument of the Fresnel integrals is not large the first term in (3.70) is negligible compared with the second, and $\theta - \alpha_0 \simeq \pi$, implying $P_2(\cos\theta) \simeq p$, so that the second term is approximately equal to (3.72); whereas when the modulus of the argument of the Fresnel integrals is large, the approximations to (3.70) and (3.72) resulting from the replacement of the Fresnel integrals by the first term of their asymptotic expansion (3.56) are seen to be identical, because

$$P_2(\cos\alpha) = p + \cos\left(\frac{\alpha-\alpha_0}{2}\right) P_1(\cos\alpha).$$

3.4. RELATION TO OTHER REPRESENTATIONS

It is instructive to relate the plane wave representation of the electromagnetic field to each of two other representations which are frequently used.

Consider, first, the expression in (2.72) for the component E_z of the two-dimensional E-polarized field generated by the current density (2.75) flowing in the surface $y = 0$. If in (2.72) the formula (2.76) is substituted for the function P, then the α integration can be effected in terms of a Hankel function, using (2.98), and the result is

$$E_z = -\tfrac{1}{4}k_0 Z_0 \int_{-\infty}^{\infty} j_z(\xi)\, H_0^{(2)}(k_0 R)\, d\xi, \tag{3.73}$$

where

$$R^2 = (x-\xi)^2 + y^2. \tag{3.74}$$

Evidently (3.73) is precisely what would have been obtained by the direct application of (2.97) to every current filament $j_z(\xi)\, d\xi$, followed by integration over all values of ξ. Together with the corresponding expression for an H-polarized field it is the two-dimensional analogue of what is probably the commonest way of writing down the vacuum field. This latter can be recovered in an entirely similar manner by substituting for P and Q from (2.111), (2.112) into (2.107). After the substitution the integration with respect to l and m can be effected by virtue of (2.120). For example, it is found that

$$H_x = -\frac{1}{4\pi} \iint_{-\infty}^{\infty} j_y(\xi,\eta)\, \frac{\partial}{\partial z}\left(\frac{e^{-ik_0R}}{R}\right) d\xi\, d\eta, \tag{3.75}$$

where

$$R^2 = (x-\xi)^2 + (y-\eta)^2 + z^2. \tag{3.76}$$

This and similar expressions for H_y and H_z can together be written

$$\mathbf{H} = -\frac{1}{4\pi}\iint_{-\infty}^{\infty} \mathbf{j}(\xi,\eta) \wedge \operatorname{grad}\left(\frac{e^{-ik_0R}}{R}\right) d\xi\, d\eta, \tag{3.77}$$

which is recognized as the standard representation of the magnetic vector as an integral over the current distribution, in the case when that distribution is a surface density of current flowing in the plane $z = 0$; it puts in evidence the fact that each element $d\xi\, d\eta$ is in effect a dipole of strength $\mathbf{j} d\xi\, d\eta$.

The other technique to be related to the plane wave representation is that based on taking the triple Fourier transform of Maxwell's equations with respect to rectangular cartesian coordinates x, y, z. In vacuum, for example, Maxwell's equations are (1.7), (1.8), with transforms

$$\mathbf{k} \wedge \mathbf{E}_k = -k_0 Z_0 \mathbf{H}_k, \tag{3.78}$$

$$\mathbf{k} \wedge \mathbf{H}_k = k_0 Y_0 \mathbf{E}_k - i\mathbf{J}_k, \tag{3.79}$$

where here

$$\mathbf{k} = k_0(l, m, n), \tag{3.80}$$

$$\mathbf{H}_k(l, m, n) = \iiint_{-\infty}^{\infty} \mathbf{H}(x, y, z)\, e^{-ik_0(lx+my+nz)}\, dx\, dy\, dz, \tag{3.81}$$

with $\mathbf{E}_k$, $\mathbf{J}_k$ similarly defined, and as before $k_0 = \omega\sqrt{\varepsilon_0\mu_0}$, $Z_0 = 1/Y_0 = \sqrt{\mu_0/\varepsilon_0}$. The algebraic equations (3.78), (3.79) are readily solved for $\mathbf{H}_k$ by taking the vector product of (3.79) with $\mathbf{k}$ and using the fact that $\mathbf{k}.\,\mathbf{H}_k = 0$; this gives

$$\mathbf{H}_k = -i\frac{\mathbf{k} \wedge \mathbf{J}_k}{k_0^2 - \mathbf{k}^2}. \tag{3.82}$$

The substitution of (3.82) into the inverse of (3.81) yields the representation

$$\mathbf{H} = -\frac{ik_0^2}{8\pi^3}\iiint_{-\infty}^{\infty} \frac{(l, m, n) \wedge \mathbf{J}_k}{1 - l^2 - m^2 - n^2}\, e^{ik_0(lx+my+nz)}\, dl\, dm\, dn. \tag{3.83}$$

The case in which the current distribution is a surface density (j_x, j_y) flowing in the plane $z = 0$ is treated by writing

$$\mathbf{J} = (j_x, j_y, 0)\,\delta(z). \tag{3.84}$$

Then $J_{kz} = 0$ and

$$(J_{kx}, J_{ky}) = \iint\limits_{-\infty}^{\infty} (j_x, j_y)\, e^{-ik_0(ln+my)}\, dx\, dy$$

$$= \frac{8\pi^2}{k_0^2}(-Q, P),$$

reintroducing the notation of (2.111), (2.112); and (3.83) becomes

$$\mathbf{H} = -\frac{i}{\pi}\iiint\limits_{-\infty}^{\infty} \frac{(-nP, -nQ, lP+mQ)}{1-l^2-m^2-n^2}\, e^{ik_0(lx+my+nz)}\, dl\, dm\, dn\,. \tag{3.85}$$

The plane wave representation can be recovered from this expression by effecting the n-integration. When $l^2 + m^2 > 1$ the poles at $n = \pm\sqrt{1-l^2-m^2}$ in the complex n-plane lie on the imaginary axis on either side of the path of integration, and when $l^2 + m^2 < 1$ the path of integration must be indented so that it passes under the pole at $-\sqrt{1-l^2-m^2}$ on the negative real axis, and above the pole at $\sqrt{1-l^2-m^2}$ on the positive real axis. Then according as $z > 0$ or $z < 0$ the path is closed by an infinite semi-circle above or below the real axis. Plainly, (2.107) is obtained.

3.5. GAIN AND SUPERGAIN

The exposition of general theory related to the plane wave spectrum representation is now complete. So far its function has been illustrated only with the very simplest examples, and the present section, which concludes the first part of the book, offers a brief discussion of a direct but rather more substantial application. Detailed applications to problems of some breadth and depth constitute the second part of the book.

In the language of aerial (antenna) theory an important feature of the radiation field associated with the plane distribution of surface current density, or with the aperture plane of a tangential field distribution, is the so-called *polar diagram* that describes the relative magnitude of the time-averaged power flux in any direction. More specifically, the concept of aerial *gain* is introduced: for radiation into a half-space this is defined as

$$G(\theta, \varphi) = 2\pi \frac{\text{power radiated per unit solid angle in direction } \theta, \varphi}{\text{total power radiated}}, \tag{3.86}$$

with the two-dimensional counterpart

$$G(\theta) = \pi \frac{\text{power radiated per unit angle in direction } \theta}{\text{total power radiated}}. \tag{3.87}$$

Thus with the representation (2.72), (2.73) it is evident from (3.4) that

$$G(\theta) = \frac{\pi\,|P(\cos\theta)|^2}{\int_0^\pi |P(\cos\alpha)|^2\,d\alpha}. \tag{3.88}$$

The corresponding expression for (3.86) can be written down from (3.15).

Highly directional radiation patterns are often required, and it may be asked how large the gain in any particular direction can be made.

The direct method of increasing the gain is simply to increase the effective excitation area of the current or aperture distribution. Take as an example the two-dimensional E-polarized field generated by the surface current density, in the plane $y = 0$,

$$\mathbf{j} = [0, 0, j(x)], \tag{3.89}$$

where

$$j(x) = \begin{cases} j_0 & \text{for } |x| < a, \\ 0 & \text{for } |x| > a, \end{cases} \tag{3.90}$$

j_0 being a constant. Then, from (2.76) (cf. (1.35), (1.36)),

$$P(\lambda) = -\frac{k_0 Z_0}{4\pi} j_0 \int_{-a}^{a} e^{ik_0 x\lambda}\,dx = -\frac{Z_0 j_0}{2\pi}\frac{\sin(k_0 a\lambda)}{\lambda}; \tag{3.91}$$

and, from (3.88),

$$G(\tfrac{1}{2}\pi) = \pi \Big/ \int_0^\pi \left[\frac{\sin(k_0 a\cos\alpha)}{k_0 a\cos\alpha}\right]^2 d\alpha. \tag{3.92}$$

Consider in turn the two extremes $k_0 a \ll 1$, $k_0 a \gg 1$. In the former the integrand in (3.92) is approximately unity, and

$$G(\tfrac{1}{2}\pi) \simeq 1. \tag{3.93}$$

To treat the latter, put $k_0 a\cos\alpha = \tau$ in the integral in (3.92) to cast it into the form

$$\frac{1}{k_0 a}\int_{-k_0 a}^{k_0 a} \frac{\sin^2\tau}{\tau^2\sqrt{1 - \tau^2/(k_0 a)^2}}\,d\tau. \tag{3.94}$$

The first order approximation for $k_0 a \gg 1$ is then seen to be

$$\frac{1}{k_0 a}\int_{-\infty}^{\infty}\frac{\sin^2\tau}{\tau^2}\,d\tau = \frac{\pi}{k_0 a}, \tag{3.95}$$

giving

$$G(\tfrac{1}{2}\pi) \simeq k_0 a. \tag{3.96}$$

Thus, for a uniform current strip, the gain in the direction normal to the plane of the strip is approximately proportional to the strip width, provided the width exceeds a wavelength.

With the recognition that there must inevitably be some restriction on the effective excitation area of a current or aperture distribution, it may further be asked to what extent the gain can be increased by arranging a suitable distribution within a given area. There is, indeed, no limit to the theoretical gain that can be achieved by ideal choice of distribution, but this result is of limited significance where practical distributions are concerned. In fact, for excitation areas whose linear dimensions are large compared with a wavelength it is difficult to improve appreciably on the *normal gain*, G_0, given by a uniform distribution. Such improvement is, however, readily achieved for small excitation areas. Gain that exceeds G_0 is called *supergain.*

To amplify these statements, consider again the current density (3.89), but now take

$$j = \begin{cases} \sum_{-N}^{N} j_n e^{\,in\pi x/a} & \text{for } |x| < a, \\ 0 & \text{for } |x| > a, \end{cases} \tag{3.97}$$

where all j_n are constants, and $j_{-n} = j_n$. Then

$$P(\lambda) = -\frac{Z_0}{2\pi}\frac{\sin(k_0 a\lambda)}{\lambda}\left[j_0 - 2k_0^2a^2\lambda^2\sum_{1}^{N}\frac{(-)^n j_n}{n^2\pi^2 - k_0^2a^2\lambda^2}\right]. \tag{3.98}$$

It is observed that

$$P(0) = -\frac{Z_0 j_0}{2\pi}k_0 a \tag{3.99}$$

is independent of $j_1, j_2, \ldots j_N$, so the gain in the direction $\theta = \frac{1}{2}\pi$ is maximized by choosing these parameters to make the total radiated power a minimum.

Suppose, for simplicity, that

$$j_1 = \eta j_0, \quad j_2 = j_3 = \cdots = j_N = 0.$$

Then (3.88) gives

$$\frac{1}{G(\frac{1}{2}\pi)} = \frac{1}{\pi}\int_0^\pi \left|\frac{\sin(k_0 a\cos\alpha)}{k_0 a\cos\alpha}\left(1+\frac{2k_0^2a^2\cos^2\alpha}{\pi^2-k_0^2a^2\cos^2\alpha}\eta\right)\right|^2 d\alpha. \tag{3.100}$$

It is easy to show that the value of η that minimizes (3.100) must be real, and that at this value of η

$$1-\frac{G_0}{G} = \frac{\left[\int_0^\pi \frac{\sin^2(k_0a\cos\alpha)}{\pi^2-k_0^2a^2\cos^2\alpha}\,d\alpha\right]^2}{\int_0^\pi \frac{\sin^2(k_0a\cos\alpha)}{\cos^2\alpha}\,d\alpha \int_0^\pi \frac{\cos^2\alpha\sin^2(k_0a\cos\alpha)}{(\pi^2-k_0^2a^2\cos^2\alpha)^2}\,d\alpha}, \tag{3.101}$$

where G_0 is the normal gain (3.92).

If $k_0a \ll 1$ approximations to the integrals in (3.101) are easily found by expanding the integrands as power series in k_0a. The first order approximation is

$$1-G_0/G \simeq \tfrac{2}{3},$$

so that

$$G \simeq 3G_0. \tag{3.102}$$

If $k_0a \gg 1$, put $k_0a\cos\alpha = \tau$ to get

$$\int_0^\pi \frac{\cos^2\alpha\sin^2(k_0a\cos\alpha)}{(\pi^2-k_0^2a^2\cos^2\alpha)^2}\,d\alpha$$

$$= \frac{1}{(k_0a)^3}\int_{-k_0a}^{k_0a} \frac{\tau^2\sin^2\tau\,d\tau}{(\pi^2-\tau^2)^2\sqrt{1-\tau^2/(k_0a)^2}}, \tag{3.103}$$

with the first order approximation

$$\frac{1}{(k_0a)^3}\int_{-\infty}^{\infty} \frac{\tau^2\sin^2\tau}{(\pi^2-\tau^2)^2}\,d\tau = \frac{\pi}{2(k_0a)^3}, \tag{3.104}$$

the integral in (3.104) being evaluated by contour integration. The same technique applied to

$$\int_0^\pi \frac{\sin^2(k_0a\cos\alpha)}{\pi^2-k_0^2a^2\cos^2\alpha}\,d\alpha \tag{3.105}$$

gives zero as the first order approximation. A more efficacious method is to write (3.105) as

$$\operatorname{Re}\frac{1}{4}\int_{-\pi}^{\pi}\frac{1-e^{-2ik_0a\cos\alpha}}{\pi^2-k_0^2a^2\cos^2\alpha}\,d\alpha. \tag{3.106}$$

Without altering the value of (3.106) the path of integration can be taken as the path L shown by the full line in Fig. 3.3. Furthermore

$$\int_L \frac{d\alpha}{\pi^2-k_0^2a^2\cos^2\alpha}$$

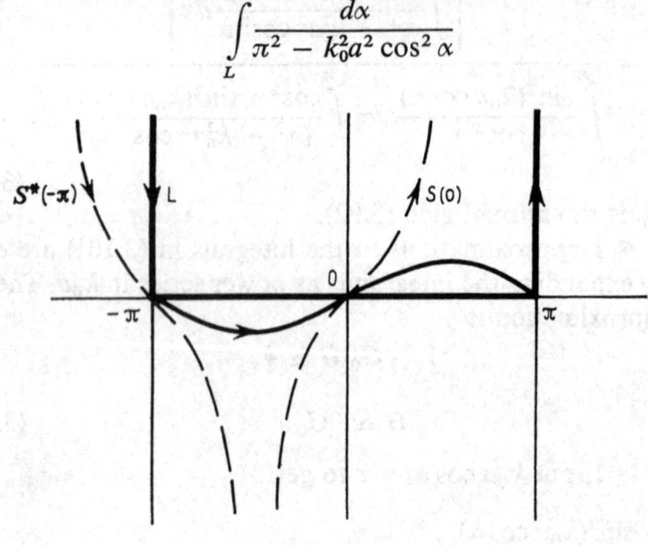

FIG. 3.3.

can be evaluated by closing the path round the poles, and is therefore pure imaginary. On the other hand, the factor $\exp(-2ik_0a \times\cos\alpha)$ in the remaining integral enables L in turn to be replaced by the paths $S^*(-\pi)$ and $S(0)$, shown dashed in Fig. 3.3; moreover, the integral over $S^*(-\pi)$ is the complex conjugate of that over $S(0)$. Hence (3.106) can be written

$$\operatorname{Re} -\frac{1}{2}\int_{S(0)}\frac{e^{-2ik_0a\cos\alpha}}{\pi^2-k_0^2a^2\cos^2\alpha}\,d\alpha, \tag{3.107}$$

and in this form the method of steepest descents is applicable when $k_0a \gg 1$. The saddle-point is at $\alpha = 0$, well away from the poles, which are close to $\alpha = \pm\frac{1}{2}\pi$; the asymptotic approximation stated in (3.39) is therefore valid, and (3.107) is approximately

$$\mathrm{Re}\,\frac{\sqrt{\tfrac{1}{2}\pi}}{k_0^2a^2-\pi^2}\,e^{\frac{1}{4}i\pi}\,\frac{e^{-2ik_0a}}{\sqrt{2k_0a}}. \tag{3.108}$$

Thus the first order approximation to (3.105) is

$$\frac{1}{2}\sqrt{\pi}\,\frac{\cos(2k_0a-\tfrac{1}{4}\pi)}{(k_0a)^{5/2}}. \tag{3.109}$$

The first integral in the denominator on the right-hand side of (3.101) has already been shown to have the approximation πk_0a. The final result is therefore

$$1-\frac{G_0}{G}\simeq\frac{1}{2\pi}\,\frac{\cos^2(2k_0a-\tfrac{1}{4}\pi)}{(k_0a)^3}, \tag{3.110}$$

so that the fractional increase of G over G_0 is very small.

Part II
APPLICATION

CHAPTER IV

DIFFRACTION BY A PLANE SCREEN

4.1. BLACK SCREEN

4.1.1. Formulation of the Problem

The phenomenon of diffraction was originally manifest in optics, long before the development of electromagnetic theory offered the chance of treating it as a boundary value problem. The typical arrangement was a source illuminating an aperture in an opaque screen; and for theoretical purposes the screen was considered to be infinitely thin and perfectly "black", the latter term implying complete absorption, without reflection, of any incident light. The concept of an infinitely thin screen that is perfectly absorbing has (in contrast to that which is perfectly reflecting) no clear cut formulation in electromagnetic theory. It does, nevertheless, lead to a useful approximate treatment; and this is comparatively simple mathematically just by virtue of having to proceed from postulated boundary *values* rather than from prescribed boundary *conditions*.

The essence of the theory of diffraction by an aperture in a thin black screen is the working hypothesis that in the aperture the field is the undisturbed field of the source (that is to say, the field that would exist there were the screen absent), whereas just behind the screen (on the other side from the source) the field is zero. This statement is not precise; the way in which it is made so in the standard electromagnetic formulation is now expounded for the special case of a plane screen, the exposition being developed in terms of the plane wave spectrum concept.

It is supposed, for the sake of definiteness, that the source of the time-harmonic field is located in the half-space $z > 0$, and that the screen occupies the whole of the plane $z = 0$ apart from a certain region (or regions) designated the *aperture*. In general it is not necessary to think of the area of the aperture as finite, nor that of the screen as infinite; and if in some specific statement it is convenient so to particularize, it should be borne in mind that there is a simple relation between the respective fields in the two cases obtained from each other by the interchange of the areas occupied by screen and aperture. This relation, commonly known as Babinet's principle, is enunciated a little further on.

The field throughout $z < 0$ (except close to the screen) is to be calculated by recognizing that it is determined by the field distribution in $z = 0$, and making an assumption about the latter of the kind just mentioned. However, it is clear from the representation of the field as an angular spectrum of plane waves, exemplified by (2.107) and (2.108) with the lower sign, that in general the specification in $z = 0$ of just two of the six components of **E**, **H** determines uniquely the spectrum functions P and Q, and hence the entire field in $z < 0$. This implies, of course, that the behaviour in $z = 0$ of the remaining four components of **E** and **H** is a consequence of the prescription of that of the two selected components, and cannot itself be arbitrarily chosen. Moreover, the question as to which two field components should be selected for the purpose of formulating the theory is a material one; for the assumption to be made about their behaviour in $z = 0$ is bound to be inexact, and different pairs will give different results. It might perhaps be thought that deviations between the results would just be of the order of the error inherent in the method, but even this is only partly true. These points are clarified in the ensuing discussion, which leads to a precise enunciation of the theory.

Any use of the components normal to the screen can be ruled out as the basis of a generally applicable theory. For if the chosen pair included E_z or H_z, or both, the legitimate adoption, for the undisturbed source field, of the particular normally incident, linearly polarized plane wave for which the selected pair was identically zero would then lead, via the proposed theory, to the absurd conclusion that the field was zero throughout the entire region $z < 0$. Attention must therefore be confined to the tangential components of **E** and **H**; and since the orientation of the x- and y-directions should have no significance in a general theory, this means the admission as possibilities of the pairs H_x, H_y and E_x, E_y.

If the pair H_x, H_y is selected, the assumption is that the field in $z \leqq 0$ is matched to values of tangential **H** in $z = 0$ that are specified in the aperture and are zero on the screen; in other words, the field is that produced by a surface electric current density whose distribution is confined to the aperture. If, on the other hand, the pair E_x, E_y is selected, the field associated with the assumed boundary values can be conceived as produced by a specified magnetic surface current density whose distribution is confined to the aperture. Each of these procedures gives, then, a readily assimilable physical picture; the field in $z < 0$ is generated by a "secondary" source distribution in the aperture.

At best, it might be expected that, in appropriate circumstances, the respective fields derived from the two procedures would be nearly the same. They are certainly not identical. For example, the assumption that H_x and H_y are zero on the shaded surface of the screen by no means implies that E_x and E_y are zero there. The final point to make is that the standard form of the theory lays down that the field be taken as half the sum of the two fields just described. The reason for this apparently somewhat arbitrary stipulation is now explained.

Evidently the exact field in $z < 0$ could be generated by the appropriate (unique) distribution in $z = 0$ of either (a) surface electric current density $\mathbf{j}$, or (b) surface magnetic current density $\mathbf{j}_m$, or (c) the superposition of electric current density $p\mathbf{j}$ and magnetic current density $(1-p)\,\mathbf{j}_m$ for any constant amplitude-phase factor p. Now surface current densities in $z = 0$ throw back into the half-space $z > 0$ the "images" of the fields they project into $z < 0$; but because $\mathbf{j}$ produces a discontinuity in tangential $\mathbf{H}$, whereas $\mathbf{j}_m$ produces one in tangential $\mathbf{E}$, the respective fields due to $\mathbf{j}$ and $\mathbf{j}_m$ in $z > 0$ are of opposite sign. The choice $p = \frac{1}{2}$ therefore specifies the unique distribution of electric and magnetic surface current densities in $z = 0$ that reproduces the given field in $z < 0$, and zero field throughout $z > 0$. It is the absence of "back" radiation as a desideratum for the concept of secondary sources that has led to the particular theoretical formulation of the diffraction problem stated above. Again, though, it is to be observed that, since in this formulation the assumed distributions of electric and magnetic surface currents individually correspond to different fields in $z < 0$, their combined field will not be exactly zero in $z > 0$. The residual field in $z > 0$ that appears in this way probably represents nothing beyond the error in the method, and has no relation to whatever small disturbance actually exists in front of the screen.

To sum up: the approximate theory of the diffraction of an incident field $\mathbf{E}^i$, $\mathbf{H}^i$ by a plane black screen assumes that the aperture carries both an electric surface current density $\mathbf{n} \wedge \mathbf{H}^i$ and a magnetic surface current density $-\mathbf{n} \wedge \mathbf{E}^i$, where $\mathbf{n}$ is the unit vector normal to the screen pointing into the half-space, behind the screen, in which the field is calculated.

Before proceeding to an explicit problem, two immediate corollaries may be noted. The first is that, for a given diffracting plane and incident field, the field behind a screen with two separate apertures, A_1, A_2, is simply the superposition of the field behind a screen with single aperture A_1 on that behind a screen with single aperture A_2. The second corollary is Babinet's principle, mentioned earlier. The diffracting plane and incident field are again supposed given, and two screens are considered that are *complementary* in the sense that they fit together like a two-piece jigsaw puzzle just to cover the entire diffracting plane. Then the principle is simply that, behind the diffracting plane, the fields pertaining to a screen and to the complementary screen, respectively, together make up the incident field that would exist there in the absence of any screen.

4.1.2. The Half-plane

One of the most important illustrations of diffraction theory is offered by the problem in which the screen is a semi-infinite plane with a straight edge.

Consider the two-dimensional case of the E-polarized plane wave

$$\mathbf{E}^i = (0, 0, 1)\, e^{ik_0 \cos(\theta - \alpha_0)} \tag{4.1}$$

incident on a thin black screen occupying the half-plane $y = 0$, $x > 0$. Here $x = r\cos\theta$, $y = r\sin\theta$, as in § 2.2.2. It is convenient to let θ run between 0 and 2π, the limiting values specifying the illuminated and shadowed faces of the screen; and to take $0 < \alpha_0 < \pi$.

Now adopt, for $y < 0$ $(\pi < \theta < 2\pi)$, the plane wave spectrum representation (2.72), (2.73); specifically

$$E_z = \int_C P(\cos\alpha)\, e^{-ik_0 r \cos(\theta + \alpha)}\, d\alpha, \tag{4.2}$$

$$H_x = -Y_0 \int_C \sin\alpha\, P(\cos\alpha)\, e^{-ik_0 r \cos(\theta + \alpha)}\, d\alpha. \tag{4.3}$$

Following the prescription of § 4.1.1, the task is to find the spectrum function $P(\cos\alpha)$ first when at $y = 0$

$$E_z = \begin{cases} 0 & \text{for} \quad x > 0, \\ e^{ik_0 x \cos\alpha_0} & \text{for} \quad x < 0, \end{cases} \tag{4.4}$$

and secondly when at $y = 0$

$$H_x = \begin{cases} 0 & \text{for} \quad x > 0, \\ -Y_0 \sin\alpha_0 \, e^{ik_0 x \cos\alpha_0} & \text{for} \quad x < 0. \end{cases} \tag{4.5}$$

The spectrum function for the problem is half the sum of the two functions thus determined.

Evidently (4.4) states

$$\int_{-\infty}^{\infty} \frac{P(\lambda)}{\sqrt{1-\lambda^2}} e^{-ik_0 x\lambda} \, d\lambda = \begin{cases} 0 & \text{for} \quad x > 0, \\ e^{ik_0\lambda_0 x} & \text{for} \; x < 0, \end{cases} \tag{4.6}$$

where it is convenient to write $\lambda_0 = \cos\alpha_0$. But the right-hand side of (4.6) is just one of the simple Fourier transform results noted in § 1.3. It follows that

$$\frac{P(\lambda)}{\sqrt{1-\lambda^2}} = \frac{1}{2\pi i} \frac{1}{\lambda + \lambda_0}, \tag{4.7}$$

where the path of integration is shown in Fig. 2.2 (λ-plane), with the addition of an indentation below the pole at $\lambda = -\lambda_0$. Thus

$$P(\cos\alpha) = \frac{1}{2\pi i} \frac{\sin\alpha}{\cos\alpha + \cos\alpha_0}. \tag{4.8}$$

Similarly (4.5) leads to

$$P(\cos\alpha) = \frac{1}{2\pi i} \frac{\sin\alpha_0}{\cos\alpha + \cos\alpha_0}. \tag{4.9}$$

The field in $y < 0$ is therefore given by (4.2), with

$$P(\cos\alpha) = \frac{1}{4\pi i} \frac{\sin\alpha + \sin\alpha_0}{\cos\alpha + \cos\alpha_0}; \tag{4.10}$$

that is, by

$$E_z = \frac{1}{4\pi i} \int_C \tan\left(\frac{\alpha + \alpha_0}{2}\right) e^{-ik_0 r \cos(\theta+\alpha)} \, d\alpha, \tag{4.11}$$

where the path of integration is indented above the pole at $\alpha = \pi - \alpha_0$, as in Fig. 4.1.

The final step is to evaluate (4.11). This need be done only for the case $k_0 r \gg 1$, since the theory is not otherwise likely to be a good approximation to any actual physical problem. The asymptotic methods of §§ 3.2, 3.3 can therefore be applied.

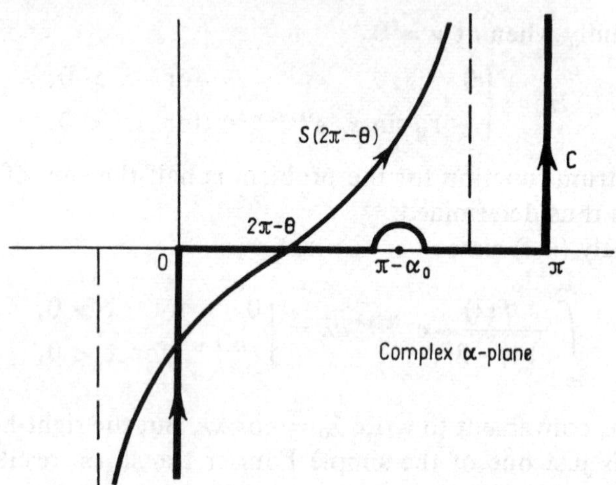

FIG. 4.1.

First, it is noted that the infinite extent of the aperture distribution has resulted in a spectrum function with poles. Now the relevant saddle-point is at $\alpha = 2\pi - \theta$ (remembering that $\pi < \theta < 2\pi$), and so allowance must be made for the saddle-point lying arbitrarily close to the pole $\alpha = \pi - \alpha_0$. The two coalesce when $\theta = \pi + \alpha_0$, and in distorting the original path of integration C to the steepest descents path $S(2\pi - \theta)$ the pole is captured or not according as to whether θ is less or greater then $\pi + \alpha_0$. The implication of this last remark is readily understood: points for which $\theta > \pi + \alpha_0$ are in the shadow of the screen, where the field is for the most part comparatively small, and points for which $\theta < \pi + \alpha_0$ are in the illuminated region, where the field is mostly not very different from the incident field; and the dominant part of the transition is provided by the contribution of the pole.

Explicitly,

$$E_z = E_z^g + E_z^d, \tag{4.12}$$

where

$$E_z^g = \begin{cases} e^{ik_0 r\cos(\theta-\alpha_0)} & \text{for } \pi < \theta < \pi + \alpha_0, \\ 0 & \text{for } \pi + \alpha_0 < \theta < 2\pi, \end{cases} \tag{4.13}$$

is the "geometrical optics" field, and

$$E_z^d = \frac{1}{4\pi i} \int\limits_{S(2\pi-\theta)} \tan\left(\frac{\alpha + \alpha_0}{2}\right) e^{-ik_0 r \cos(\theta+\alpha)}\, d\alpha \tag{4.14}$$

is the "diffracted" field. Now (4.14) is of the form (3.67), (3.71), with $2\pi - \theta$ replacing θ, $2\pi - \alpha_0$ replacing α_0, and

$$P_2(\cos\alpha) = -\frac{1}{4\pi i} \sin\left(\frac{\alpha + \alpha_0}{2}\right). \tag{4.15}$$

Hence, from (3.72),

$$E_z^d = \mp \frac{e^{\frac{1}{4}i\pi}}{\sqrt{\pi}} \sin\left(\frac{\theta - \alpha_0}{2}\right) e^{-ik_0 r} F\left[\pm \sqrt{2k_0 r} \cos\left(\frac{\theta - \alpha_0}{2}\right)\right], \tag{4.16}$$

with the upper/lower sign for $\theta \lessgtr \pi + \alpha_0$.

Since $F(0) = \frac{1}{2}\sqrt{\pi} \exp(-\frac{1}{4}i\pi)$ it is immediately confirmed that the discontinuity in (4.16) at $\theta = \pi + \alpha_0$ balances that in (4.13). In the vicinity of $\theta = \pi + \alpha_0$ (4.16) is comparable with the incident field, and when superposed on it gives rise to the well-known "fringes" in the illuminated region. In the shadow region (4.16) is the only term contributing to the field, which thus falls off monotonically, for a fixed value of r, or of y, as θ increases from $\pi + \alpha_0$.

The asymptotic expansion (3.56) of the Fresnel integral can be used when, as stated in (3.66),

$$\left|\sqrt{2k_0 r} \cos\left(\frac{\theta - \alpha_0}{2}\right)\right| \gg 1. \tag{4.17}$$

From the first term of the expansion the approximation to (4.16) is

$$E_z^d = -\frac{e^{-\frac{1}{4}i\pi}}{2\sqrt{2\pi}} \tan\left(\frac{\theta - \alpha_0}{2}\right) \frac{e^{-ik_0 r}}{\sqrt{k_0 r}}, \tag{4.18}$$

which is, of course, the result that is obtained on replacing the entire factor $\tan[\frac{1}{2}(\alpha + \alpha_0)]$ in the integrand in (4.14) by its value at the saddle-point $\alpha = 2\pi - \theta$ (cf. (3.39)). Equation (4.18) is the "edge-wave" approximation, and from (4.17) is valid outside some parabola

$$r\cos(\theta - \alpha_0) + r = \text{constant}, \tag{4.19}$$

whose focus is at the diffracting edge and whose axis is along the shadow line of geometrical optics.

Finally it may be noted that since θ enters the solution only in the combination $\theta - \alpha_0$, diffraction on this theory is purely an edge effect.

4.1.3. The Slit

A brief discussion of the two-dimensional problem of diffraction by a parallel sided slit in a black screen now follows. The object here is to illustrate the part played by the finite width of the aperture.

It is supposed, then, that black screens occupy the half-planes $y = 0, x > a$ and $y = 0, x < -a$. The incident field is again taken to be the plane wave (4.1), with $x = r\cos\theta$, $y = r\sin\theta$; but to simplify the algebra only the case of normal incidence, $\alpha_0 = \frac{1}{2}\pi$, is treated explicitly.

It is, of course, possible to write down the answer immediately in terms of two expressions like (4.16), since the field on the far side of the screen is obviously just the difference between the respective fields of two half-plane problems. However, the form of solution which this procedure gives does tend to obscure the fact that at sufficiently great distances from the slit the field can be represented in all directions as a (cylindrical) radiation field, there being no singularities in the plane wave spectrum function. It is instructive, therefore, to consider the direct approach.

With the help of the Fourier transforms (1.35), (1.36) the stages of § 4.1.2 that led to (4.10) are now easily seen, with $\alpha_0 = \frac{1}{2}\pi$, to yield the spectrum function (cf. (3.91))

$$P(\cos\alpha) = \frac{1}{2\pi}(1 + \sin\alpha)\frac{\sin(k_0 a\cos\alpha)}{\cos\alpha}. \tag{4.20}$$

Since (4.20) is free of singularities, the representation (4.2) can certainly be approximated, without restriction on θ, by the asymptotic form corresponding to (3.39), provided only that $k_0 r$ is sufficiently large. The outcome is

$$E_z \sim \frac{e^{\frac{1}{4}i\pi}}{\sqrt{2\pi}}(1 - \sin\theta)\frac{\sin(k_0 a\cos\theta)}{\cos\theta}\frac{e^{-ik_0 r}}{\sqrt{k_0 r}}. \tag{4.21}$$

On the other hand, the expansion of $P(\cos\alpha)$, which is the basis of the derivation of the asymptotic development, as shown in § 3.2.2, now depends on $k_0 a$; and it appears that the inequality required to validate (4.21) is

$$k_0 r \gg (k_0 a)^2. \tag{4.22}$$

With $k_0 a \gg 1$, (4.22) is a much more severe restriction on the proximity to the slit of the inner boundary of the radiation region than the more obvious conditions $k_0 r \gg 1$ and $r \gg a$. The ratio of the square of a characteristic linear dimension of a diffracting object to the wavelength is known as the *Rayleigh distance*; and as is seen here in a special case, the radiation field is only developed, for objects with dimensions exceeding a wavelength, at distances greater than the Rayleigh distance.

What if $k_0 r \gg 1$, but the inequality (4.22) does not hold? In that case help must be sought from the Fresnel integral form of the solution. A certain amount of algebra is involved in distinguishing the various cases, and here it is merely pointed out that (4.22) is effectively the condition for points of observation at all values of x in the region $|x| < a$ to be sufficiently far from the slit for the arguments of both Fresnel integrals to be small.

4.2. PERFECTLY CONDUCTING SCREEN

4.2.1. Babinet's Principle and the Cross-section Theorem

The theory of diffraction by an infinitely thin screen presents a well-defined boundary value problem if the screen can be assumed to be perfectly conducting. This case is now considered.

Two exact results of a rather general kind are available; one is an analogue of Babinet's principle, the other a relation between the transmission or scattering cross-section and the value of the radiation field in the forward direction. These are described in turn.

When an electromagnetic field $\mathbf{E}^i$, $\mathbf{H}^i$ is incident on an infinitely thin, perfectly conducting, plane screen, surface electric currents are induced in the screen, and it is useful to think of the total field as the incident field together with the "scattered" field that is generated by the induced currents. Specifically,

$$\mathbf{E} = \mathbf{E}^i + \mathbf{E}^s, \quad \mathbf{H} = \mathbf{H}^i + \mathbf{H}^s. \tag{4.23}$$

It is emphasized that the incident field is the field that would be produced by the specified source, were that source alone in unbounded empty space. If, for example, a source in $z > 0$ were confronted by a perfectly conducting sheet filling the entire plane $z = 0$, then according to the viewpoint expressed by (4.23), the reason why there is no field in $z < 0$ is because in that region the scattered field is precisely the negative of the incident field; and this, of course, by virtue of the symmetry of the scattered field as discussed in § 2.2.1, is consistent with the fact that the scattered field in $z > 0$ represents the image in $z = 0$ of the given source.

Let the aperture region in a plane screen be designated A and the perfectly conducting region S, so that A and S together make up the entire plane. Then the fundamental boundary condition is

$$\mathbf{E}_t^i + \mathbf{E}_t^s = 0 \quad \text{on} \quad S, \tag{4.24}$$

where the suffix t denotes the tangential component. It is also most useful to state explicitly

$$\mathbf{H}_t^s = 0 \quad \text{on} \quad A. \tag{4.25}$$

Admittedly (4.25) is only a particular expression of the type of symmetry possessed by the scattered field; saying, indeed, just that the aperture carries no surface current. On the other hand, in conjunction with (4.24) and the outgoing behaviour of the waves, it does determine completely the field in the half-space behind the screen; and in this way the problem is presented as a half-space boundary value problem.

Consider, next, the complementary screen, in which A designates the conducting region, and S the aperture region. The boundary conditions are then

$$\mathbf{E}_t' = 0 \quad \text{on} \quad A, \tag{4.26}$$

$$\mathbf{H}_t' - \mathbf{H}_t'^i = 0 \quad \text{on} \quad S, \tag{4.27}$$

where the dash is used to distinguish this field from the previous one, and the conditions have been written in terms of the total field $\mathbf{E}'$, $\mathbf{H}'$ for a reason which will be apparent in a moment.

Suppose, finally, that the incident fields in the two cases are connected by the relation

$$\mathbf{H}'^i = Y_0 \mathbf{E}^i. \tag{4.28}$$

Then a comparison of (4.26) with (4.25), and (4.27) with (4.24), makes it evident that, in the half-space behind the screen, where the total as well as the scattered field is outgoing,

$$\mathbf{H}' = -Y_0 \mathbf{E}^s, \tag{4.29}$$

or equivalently

$$\mathbf{E} + Z_0 \mathbf{H}' = \mathbf{E}^i. \tag{4.30}$$

The close connection between (4.30) and the statement of Babinet's principle at the end of § 4.1.1 hardly needs pointing out, except perhaps to stress that whereas the theory behind the latter rests on assumptions that can only be approximately correct, the former is derived from an exact treatment based on Maxwell's equations and well defined boundary conditions.

The second general theorem is now considered, that relating to the cross-section. The transmission cross-section σ_T of an aperture of finite area is defined as follows: an incident monochromatic, homogeneous, plane wave is specified, and σ_T is the ratio of the time-averaged power transmitted through the aperture to the time-averaged power carried by the incident wave alone across unit area normal to its direction of propagation. The scattering cross-section σ_S of a plane conducting sheet of finite area is similarly defined: with plane wave incidence, σ_S is the ratio of the time-averaged *scattered* power to the time-averaged power flux density in the incident wave. The latter definition is, of course, applicable to *any* finite scattering obstacle; the case of an infinitely thin, perfectly conducting, plane sheet is special in that power is radiated equally on either side of the sheet, so that $\frac{1}{2}\sigma_S$ can be calculated by considering only the field in the half-space on one side of the plane of the sheet.

To show most simply the nature of the theorem and its proof, consider the E-polarized plane wave

$$\mathbf{E}^i = (0, 0, 1)\, e^{ik_0 r \cos(\theta - \alpha_0)} \tag{4.31}$$

incident on a perfectly conducting screen which lies in the plane $y = 0$ and contains a slit (or slits) of finite width, with infinite edges parallel to the z-axis. Then the time-averaged power transmitted through the slit is

$$\operatorname{Re} -\tfrac{1}{2}\int_{-\infty}^{\infty} E_z H_x^*\, dx, \tag{4.32}$$

evaluated at any given $y \leqq 0$. If y is taken zero, (4.32) can be written

$$\operatorname{Re} \tfrac{1}{2} Y_0 \sin\alpha_0 \int_{\text{aperture}} E_z\, e^{-ik_0 x \cos\alpha_0}\, dx, \tag{4.33}$$

since E_z is zero on the conductor and $H_x = H_x^i$ in the aperture. But when the field in $y < 0$ is expressed in terms of the spectrum function $P(\cos\alpha)$, as in (4.2), then

$$\frac{P(\lambda)}{\sqrt{1-\lambda^2}} = \frac{k_0}{2\pi}\int E_z\, e^{ik_0 x\lambda}\, dx, \tag{4.34}$$

and (4.33) is seen to be proportional to the real part of $P(-\cos\alpha_0)$. In fact, division by the time-averaged power flux density in the incident wave, $\frac{1}{2}Y_0$, gives

$$\sigma_T = \mathrm{Re}\,\frac{2\pi}{k_0}\,P(-\cos\alpha_0). \tag{4.35}$$

Now (4.35) can in turn be expressed in terms of the radiation field in the forward direction, that is, in the direction of propagation of the incident wave, $\theta = \pi + \alpha_0$. For since the aperture is of finite width, the far field at $\theta = \pi + \alpha_0$ is given by

$$E_z = F\frac{e^{-ik_0 r}}{\sqrt{r}}, \tag{4.36}$$

where

$$F = \sqrt{2\pi/k_0}\,e^{\frac{1}{4}i\pi}\,P(-\cos\alpha_0). \tag{4.37}$$

Hence

$$\sigma_T = \mathrm{Im}\,e^{\frac{1}{4}i\pi}\sqrt{2\pi/k_0}\,F. \tag{4.38}$$

The content of the cross-section theorem is represented by the connection shown in (4.36) and (4.38) between the cross-section and the amplitude of the far field in the forward direction.

It is easy to confirm, in the two-dimensional case, that the result just proved does in fact apply to the incidence of any homogeneous, linearly polarized plane wave with electric vector $\mathbf{E}^i$ of unit amplitude and arbitrary direction. Formula (4.38) is of general validity, it being only necessary to interpret F as the quantity which identifies the right-hand side of (4.36) with the component of $\mathbf{E}$, in the far field, that is parallel to $\mathbf{E}^i$.

The three-dimensional case follows just the same lines. For a perfectly conducting screen in the plane $z = 0$, with aperture A of finite area, the time-averaged transmitted power is

$$\mathrm{Re}\,\tfrac{1}{2}\iint\limits_A (-E_x H_y^{i*} + E_y H_x^{i*})\,dx\,dy. \tag{4.39}$$

Without loss of generality the incident wave can be written

$$\begin{aligned}\mathbf{E}^i &= (\cos\gamma,\ -n\sin\gamma,\ m\sin\gamma)\,e^{ik_0(my+nz)},\\ \mathbf{H}^i &= Y_0(-\sin\gamma,\ -n\cos\gamma,\ m\cos\gamma)\,e^{ik_0(my+nz)},\end{aligned} \tag{4.40}$$

where $x = 0$ is chosen as the plane of incidence, γ is the arbitrary angle which $\mathbf{E}^i$ makes with the x-axis, and $n = \sqrt{1-m^2}$. Then

(4.39) at $z = 0$ is

$$\operatorname{Re} \tfrac{1}{2} Y_0 \iint_A (E_x n \cos\gamma - E_y \sin\gamma)\, e^{-ik_0 my}\, dx\, dy. \qquad (4.41)$$

The representations (2.107), (2.108) are now used; in particular, the inverse of (2.108) with $z = 0$. Evidently (4.41) can be expressed in terms of $P(0, m)$ and $Q(0, m)$; and after division by $\tfrac{1}{2}Y_0$ to get the transmission cross-section, it appears that

$$\sigma_T = \operatorname{Re} \frac{4\pi^2 Z_0}{k_0^2} (nP \sin\gamma + Q \cos\gamma). \qquad (4.42)$$

But from (3.22) the far field in the forward direction, $\varphi = 3\pi/2$, $\cos\theta = -n$, is

$$(E_\theta, E_\varphi) \sim Z_0(H_\varphi, -H_\theta) \sim 2\pi i Z_0(-nP, Q) \frac{e^{-ik_0 r}}{k_0 r},$$

so that the far field component of $\mathbf{E}$ parallel to $\mathbf{E}^i$ is

$$E_{||} \sim F \frac{e^{-ik_0 r}}{r}, \qquad (4.43)$$

where

$$F = 2\pi i \frac{Z_0}{k_0} (nP \sin\gamma + Q \cos\gamma). \qquad (4.44)$$

Thus (4.42) states that

$$\sigma_T = \operatorname{Im}(2\pi/k_0)\, F. \qquad (4.45)$$

It only remains to observe that virtually identical results apply for the scattering cross-section, σ_S, when a homogeneous, linearly polarized plane wave, with electric vector of unit amplitude, falls on a plane conducting sheet of finite area. It is easy to see from the exact form of Babinet's principle, expressed by (4.30), that, in the two-dimensional case,

$$\sigma_S = -\operatorname{Im} e^{\frac{1}{4}i\pi}\, 2\sqrt{2\pi/k_0}\, F, \qquad (4.46)$$

and in the three-dimensional case

$$\sigma_S = -\operatorname{Im}(4\pi/k_0)\, F, \qquad (4.47)$$

where F now specifies the complex amplitude in the radiation region of the component parallel to $\mathbf{E}^i$ of the electric vector $\mathbf{E}^s$ of the *scattered* field, and the reason for the extra factor 2 has already been mentioned.

4.2.2. The Half-plane

The specification of the problem now considered differs from that of the problem treated in § 4.1.2 only by having a perfectly conducting rather than a perfectly absorbing screen.

Electric surface currents parallel to the z-axis are induced in the half-plane $y = 0$, $x > 0$ through the action of the incident plane wave (4.1). The scattered field $\mathbf{E}^s$, $\mathbf{H}^s$ is represented as an angular spectrum of plane waves precisely as in (2.72), (2.73). It is recalled that in this representation the continuity of E_z across $y = 0$ is automatically ensured. The boundary conditions from which the spectrum function $P(\cos\alpha)$ is to be determined are therefore (4.24) and (4.25). In the present context these are, in reverse order,

$$\int_{-\infty}^{\infty} P(\lambda)\, e^{-ik_0x\lambda}\, d\lambda = 0, \quad \text{for} \quad x < 0, \tag{4.48}$$

$$\int_{-\infty}^{\infty} \frac{P(\lambda)}{\sqrt{1-\lambda^2}}\, e^{-ik_0x\lambda}\, d\lambda = -e^{ik_0x\lambda_0}, \quad \text{for} \quad x > 0, \tag{4.49}$$

where $\lambda_0 = \cos\alpha_0$, and the path of integration avoids the branch-points at $\lambda = \pm 1$ in the way shown in Fig. 2.2.

It thus appears that instead of a single integral equation for $P(\lambda)$, which is effectively what characterized the problem of § 4.1.2, there are two integral equations, holding for different ranges of x. Such *dual* integral equations can be solved when the ranges are $x < 0$ and $x > 0$; and here, in fact, in contrast to most cases, the solution is very simple because of the simplicity of the functions involved.

Keeping close to the ideas of § 1.3 and § 4.1.2, it is noted that, if $P(\lambda)$ is of algebraic growth at infinity in the complex λ plane, then the paths of integration in (4.48) and (4.49) can be closed by infinite semicircles in the upper and lower half-planes respectively. The solution of the equations therefore requires only that $P(\lambda)$ be free of singularities in the upper half-plane, and that $P(\lambda)/\sqrt{1-\lambda^2}$ have a simple pole at $\lambda = -\lambda_0$, which the path of integration avoids by being indented to pass above it, and be otherwise free of singularities in the lower half-plane.

To give formal mathematical expression to these statements it is convenient to introduce the following notation: functions free of singularities and zeros in the half of the complex λ-plane above the path of integration (the upper half-plane), and of algebraic growth at infinity therein, are called U functions, and are written $U(\lambda)$; and those free of singularities and zeros in the lower half-plane, and of algebraic growth at infinity therein, are called L functions, and are written $L(\lambda)$.

Then a solution of (4.48) is

$$P(\lambda) = U(\lambda), \tag{4.50}$$

and a solution of (4.49) is

$$\frac{P(\lambda)}{\sqrt{1-\lambda^2}} = \frac{1}{2\pi i}\,\frac{L(\lambda)}{L(-\lambda_0)}\,\frac{1}{\lambda+\lambda_0}. \tag{4.51}$$

All that remains, therefore, is to find functions $U(\lambda)$ and $L(\lambda)$ to satisfy the relation

$$\frac{1}{\sqrt{1-\lambda^2}} = \frac{L(\lambda)}{L(-\lambda_0)}\,\frac{1}{2\pi i U(\lambda)\,(\lambda+\lambda_0)}. \tag{4.52}$$

Evidently (4.52) requires the explicit expression of $1/\sqrt{1-\lambda^2}$ as the product of a U function and an L function; and with Fig. 2.2 in mind the two functions are seen to be $1/\sqrt{1-\lambda}$ and $1/\sqrt{1+\lambda}$ respectively. Now $1/U(\lambda)$ and $1/(\lambda+\lambda_0)$ are both U functions, so that (4.52) implies

$$\frac{L(\lambda)}{L(-\lambda_0)} = \frac{\sqrt{1-\lambda_0}}{\sqrt{1+\lambda}}, \tag{4.53}$$

where the constant factor on the right-hand side is fixed by the fact that the left-hand side is unity when $\lambda = -\lambda_0$. Thus

$$P(\lambda) = \frac{1}{2\pi i}\,\frac{\sqrt{1-\lambda_0}\,\sqrt{1-\lambda}}{\lambda+\lambda_0}, \tag{4.54}$$

or equivalently

$$\begin{aligned} P(\cos\alpha) &= \frac{1}{\pi i}\,\frac{\sin(\tfrac{1}{2}\alpha_0)\sin(\tfrac{1}{2}\alpha)}{\cos\alpha+\cos\alpha_0} \\ &= \frac{1}{4\pi i}\left[\sec\left(\frac{\alpha+\alpha_0}{2}\right) - \sec\left(\frac{\alpha-\alpha_0}{2}\right)\right]. \end{aligned} \tag{4.55}$$

Reference to (3.65) now reveals the delightful result that the exact solution can be expressed in terms of Fresnel integrals. The details are easily worked out. The spectrum (4.55) is substituted into (2.72), (2.73) to give the scattered field, and the path C is distorted into the path of steepest descents, $S(\theta)$ for $y > 0$, $S(2\pi - \theta)$ for $y < 0$, with due allowance made for possible capture of the poles. With the help of (3.44) it is found that the electric vector of the *total* field can be written

$$E_z = \frac{e^{\frac{1}{4}i\pi}}{\sqrt{\pi}} \left\{ F\left[-\sqrt{2k_0 r} \cos\left(\frac{\theta - \alpha_0}{2}\right)\right] - F\left[-\sqrt{2k_0 r} \cos\left(\frac{\theta + \alpha_0}{2}\right)\right]\right\} e^{-ik_0 r}. \tag{4.56}$$

From (4.56) various approximate forms of the field can be derived, depending on whether $k_0 r$ is large or small and whether θ is or is not close to either of $\pi \pm \alpha_0$. Remarks analogous to those made in the paragraphs of § 4.1.2 following (4.16) are applicable, save for the fact that θ does not enter (4.56) solely in the combination $\theta - \alpha_0$.

The corresponding problem when the incident plane wave is H-polarized can be solved in a virtually identical manner; alternatively the solution can be deduced from that just obtained by appeal to the exact form of Babinet's principle. The magnetic vector of the total field turns out to be

$$H_z = \frac{e^{\frac{1}{4}i\pi}}{\sqrt{\pi}} \left\{ F\left[-\sqrt{2k_0 r} \cos\left(\frac{\theta - \alpha_0}{2}\right)\right] + F\left[-\sqrt{2k_0 r} \cos\left(\frac{\theta + \alpha_0}{2}\right)\right]\right\} e^{-ik_0 r}, \tag{4.57}$$

which differs from (4.56) only in the sign in front of the second Fresnel integral.

4.2.3. The Wide Slit

Diffraction by a two-dimensional slit in a perfectly conducting screen presents a vastly more difficult problem for exact solution than the half-plane. Recently some progress has been made in obtaining a solution in closed form, but the analysis is abstruse. Here the discussion is limited to approximate treatments of the two complementary cases in which the slit width is respectively much greater or much less than a wavelength.

Suppose that the two-dimensional, H-polarized plane wave specified by

$$H_z^i = e^{ik_0 r \cos(\theta - \alpha_0)} \tag{4.58}$$

is incident on perfectly conducting half-planes occupying $y = 0$, $x > 0$ and $y = 0$, $x < -d$; and assume without loss of generality that $0 \leqq \alpha_0 \leqq \frac{1}{2}\pi$. In the case $k_0 d \gg 1$ it is to be expected that, if α_0 is not too small, the current distribution in each half-plane will be approximated by that which would exist were the other half-plane absent. The aim, therefore, is to cast the analysis into a form in which this zero order approximation appears naturally, and from which successively better approximations can be obtained by some sort of iteration.

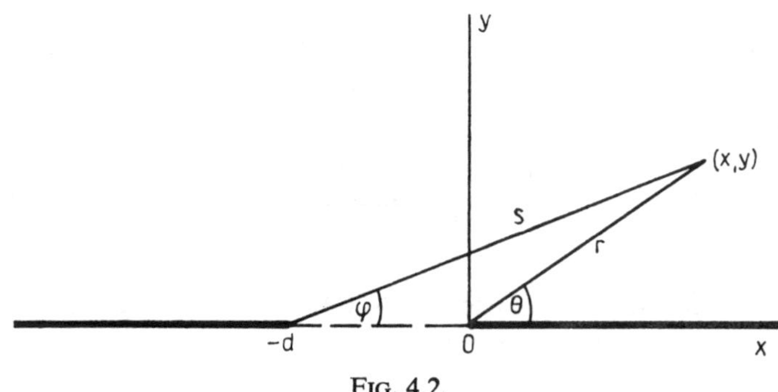

FIG. 4.2.

To this end, let r, θ and s, φ be polar coordinates measured from the respective edges of the half-planes, as illustrated in Fig. 4.2, and write the field scattered by the right half-plane as

$$H_z^{s1} = \pm \int_C P_1(\cos\alpha)\, e^{-ik_0 r \cos(\theta \mp \alpha)}\, d\alpha, \tag{4.59}$$

and that scattered by the left half-plane as

$$H_z^{s2} = \pm \int_C P_2(\cos\alpha)\, e^{ik_0 s \cos(\varphi \pm \alpha)}\, d\alpha, \tag{4.60}$$

where in each case the upper/lower sign applies for $y \gtrless 0$. Since H_z^{s1} must be zero for $\theta = \pi$, and H_z^{s2} must be zero for $\varphi = 0$, both $P_1(\lambda)/\sqrt{1-\lambda^2}$ and $P_2(\lambda)/\sqrt{1-\lambda^2}$ are taken to be U functions. The complete scattered field is the superposition of (4.59) and (4.60).

The boundary condition that the component E_x of the *total* field vanish on $y = 0$, $x > 0$ gives

$$\int_{-\infty}^{\infty} [P_1(\lambda)\, e^{-ik_0 r\lambda} + P_2(\lambda)\, e^{ik_0(d+r)\lambda}]\, d\lambda = \sqrt{1-\lambda_0^2}\, e^{ik_0 r\lambda_0}, \quad (4.61)$$

where $\lambda_0 = \cos\alpha_0$; and the same condition on $y = 0$, $x < -d$ gives

$$\int_{-\infty}^{\infty} [P_1(\lambda)\, e^{ik(s+d)\lambda} + P_2(\lambda)\, e^{-ik_0 s\lambda}]\, d\lambda = \sqrt{1-\lambda_0^2}\, e^{-ik_0(s+d)\lambda_0}. \quad (4.62)$$

These simultaneous integral equations for P_1 and P_2 can be cast into a form suitable for solution by successive approximations when $k_0 d \gg 1$ by the following device. Write

$$\int_{-\infty}^{\infty} P_2(\lambda)\, e^{ik_0(d+r)\lambda}\, d\lambda = -\int_{-\infty}^{\infty} Q_2(\lambda)\, e^{-ik_0 r\lambda}\, d\lambda, \quad (4.63)$$

where

$$Q_2(\lambda) = \frac{\sqrt{1+\lambda}}{2\pi i} \int_{-\infty}^{\infty} \frac{P_2(\mu)}{\sqrt{1-\mu}\,(\lambda+\mu)}\, e^{ik_0 d\mu}\, d\mu, \quad (4.64)$$

and the path of integration in (4.64) passes above the pole at $\mu = -\lambda$. Then (4.61) appears as

$$\int_{-\infty}^{\infty} [P_1(\lambda) - Q_2(\lambda)]\, e^{-ik_0 r\lambda}\, d\lambda = \sqrt{1-\lambda_0^2}\, e^{ik_0 r\lambda_0}. \quad (4.65)$$

But (4.65) is precisely the equation that describes the correspon-ing isolated half-plane problem, with $P_1(\lambda) - Q_2(\lambda)$ replacing $P_1(\lambda)$; for, by design, $Q_2(\lambda)/\sqrt{1-\lambda^2}$ is also a U function. The solution is (cf. (4.54))

$$P_1(\lambda) - Q_2(\lambda) = -\frac{1}{2\pi i}\, \frac{\sqrt{1+\lambda_0}\,\sqrt{1+\lambda}}{\lambda+\lambda_0}. \quad (4.66)$$

Thus (4.61) yields

$$P_1(\lambda) = -\frac{1}{2\pi i}\, \frac{\sqrt{1+\lambda_0}\,\sqrt{1+\lambda}}{\lambda+\lambda_0} + \frac{\sqrt{1+\lambda}}{2\pi i} \int_{-\infty}^{\infty} \frac{P_2(\mu)}{\sqrt{1-\mu}\,(\mu+\lambda)}\, e^{ik_0 d\mu}\, d\mu; \quad (4.67)$$

and similarly (4.62) yields

$$P_2(\lambda) = -\frac{1}{2\pi i}\frac{\sqrt{1-\lambda_0}\sqrt{1+\lambda}}{\lambda-\lambda_0}e^{-ik_0 d\lambda_0} + \frac{\sqrt{1+\lambda}}{2\pi i}\int_{-\infty}^{\infty}\frac{P_1(\mu)}{\sqrt{1-\mu}\,(\mu+\lambda)}e^{ik_0 d\mu}\,d\mu. \tag{4.68}$$

Equations (4.67) and (4.68) are in the required form. Apart from the case of nearly grazing incidence ($\lambda_0 \simeq 1$), the first terms on the right-hand side of each equation give the zero order approximations, that is, the isolated half-plane solutions. Correction terms are then obtained by substituting these zero order approximations for P_1 and P_2 into the integrands in the second terms, and it only remains to evaluate the resulting integrals.

Consider the integral in question in (4.68). Leaving a multiplying factor to be reinstated later, it is

$$\int_{-\infty}^{\infty}\frac{\sqrt{1+\mu}}{\sqrt{1-\mu}\,(\mu+\lambda)(\mu+\lambda_0)}e^{ik_0 d\mu}\,d\mu, \tag{4.69}$$

where the path of integration is shown in Fig. 4.3. The branch cut through $\mu = -1$ can be taken parallel to the positive imaginary axis, and the path of integration can be wrapped round it. If, then,

$$\mu = -1 + i\tau^2,$$

the new integration variable τ runs from $-\infty$ to ∞, and (4.69) is

$$-2e^{-\frac{1}{4}i\pi}e^{-ik_0 d}\int_{-\infty}^{\infty}\frac{\tau^2\,e^{-k_0 d\tau^2}}{\sqrt{2-i\tau^2}\,(1-\lambda-i\tau^2)(1-\lambda_0-i\tau^2)}d\tau. \tag{4.70}$$

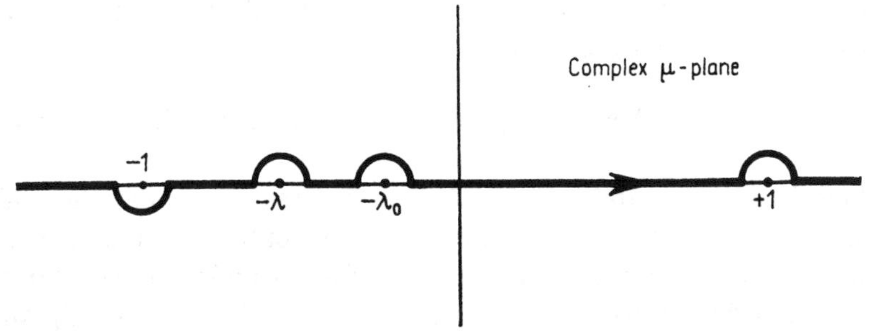

FIG. 4.3.

With $k_0 d \gg 1$, and provided λ and λ_0 are not too near unity, the asymptotic expansion in inverse powers of $k_0 d$ may be written down by developing the non-exponential part of the integrand as a power series in τ^2 and integrating term by term (see § 3.2.2). The first term gives the approximation

$$-\sqrt{\tfrac{1}{2}\pi}\, e^{-\frac{1}{4}i\pi} \frac{1}{(1-\lambda_0)(1-\lambda)} \frac{e^{-ik_0 d}}{(k_0 d)^{3/2}}. \tag{4.71}$$

In this way (4.67) and (4.68) yield the approximations

$$P_1(\lambda) = -\frac{1}{2\pi i} \frac{\sqrt{1+\lambda_0}\sqrt{1+\lambda}}{\lambda+\lambda_0} - \frac{e^{-\frac{1}{4}i\pi}}{4\pi\sqrt{2\pi}} e^{-ik_0 d\lambda_0} \frac{\sqrt{1-\lambda_0}\sqrt{1+\lambda}}{(1+\lambda_0)(1-\lambda)} \frac{e^{-ik_0 d}}{(k_0 d)^{3/2}}, \tag{4.72}$$

$$P_2(\lambda) = -\frac{1}{2\pi i} \frac{\sqrt{1-\lambda_0}\sqrt{1+\lambda}}{\lambda-\lambda_0} e^{-ik_0 d\lambda_0} - \frac{e^{-\frac{1}{4}i\pi}}{4\pi\sqrt{2\pi}} \frac{\sqrt{1+\lambda_0}\sqrt{1+\lambda}}{(1-\lambda_0)(1-\lambda)} \frac{e^{-ik_0 d}}{(k_0 d)^{3/2}}, \tag{4.73}$$

valid when $k_0 d \gg 1$, provided λ and λ_0 are not too near unity. The analysis also requires that $1+\lambda_0$ be not too near zero, but this is inherent in the assumption that α_0 is between 0 and $\frac{1}{2}\pi$.

Now the magnetic vector of the complete scattered field, obtained by summing (4.59) and (4.60), can be written

$$H_z^s = \pm \int_C [P_1(\cos\alpha) + P_2(-\cos\alpha)\, e^{-ik_0 d\cos\alpha}]\, e^{-ik_0 r\cos(\theta\mp\alpha)}\, d\alpha. \tag{4.74}$$

The expression for the transmission cross-section of the slit corresponding to the approximations (4.72), (4.73) can therefore be written down from the formula

$$\sigma_T = \operatorname{Re} -\frac{2\pi}{k_0} [P_1(-\lambda_0) + P_2(\lambda_0)\, e^{ik_0 d\lambda_0}], \tag{4.75}$$

which is the form that (4.35) takes in the present context, the leading minus sign arising solely from the convention adopted for the plane wave spectrum representations for E- and H-polarization. The first terms in (4.72) and (4.73) diverge as λ tends to $-\lambda_0$ and λ_0 respectively, but the combination in (4.75) tends to a limit. The result is

$$\sigma_T/d = \sin\alpha_0 + \frac{1}{2\sqrt{2\pi}}\left\{\frac{(1-\cos\alpha_0)\cos[k_0d(1+\cos\alpha_0)+\frac{1}{4}\pi]}{(1+\cos\alpha_0)^2}\right.$$
$$\left.+\frac{(1+\cos\alpha_0)\cos[k_0d(1-\cos\alpha_0)+\frac{1}{4}\pi]}{(1-\cos\alpha_0)^2}\right\}\frac{1}{(k_0d)^{5/2}}, \qquad (4.76)$$

which gives the first correction to the geometrical cross-section $d\sin\alpha_0$. In particular, for normal incidence, $\alpha_0 = \frac{1}{2}\pi$,

$$\sigma_T/d = 1 + \frac{1}{\sqrt{2\pi}}\frac{\cos(k_0d+\frac{1}{4}\pi)}{(k_0d)^{5/2}}. \qquad (4.77)$$

The failure of (4.76) at near grazing incidence, $\alpha_0 \sim 0$, is apparent from the nature of the denominator of the last term. This case is somewhat more complicated, for the following reason. As α_0 decreases towards zero, the current distribution in the right half-plane remains nearly the same as that which would be present were the half-plane isolated; indeed, gets closer to it, since the disturbing field it experiences due to the presence of the left half-plane becomes more nearly back scatter from a thin screen viewed end on. On the other hand, the current distribution in the left half-plane departs markedly from that which would be present were that half-plane isolated; for the disturbing field it experiences due to the presence of the right half-plane is more nearly forward scatter, which is comparable in magnitude to the incident wave. Putting it more quantitatively, the previous calculation will begin to fail when α_0 becomes so small that the edge of the left half-plane falls inside the parabola described by (4.19).

What does this mean analytically? The lowest order approximation suffices to show the change in character of the solution, and for this the integral term in (4.67) is dropped, leaving the isolated half-plane result for $P_1(\lambda)$. This is then fed into the integral in (4.68), giving, apart from a multiplying factor, the result (4.69), and in turn (4.70). However, since now it is presumed that λ_0 is near unity, the subsequent evaluation cannot follow the simple step that led from (4.70) to (4.71). For the cross-section calculation, in fact, λ is equated to λ_0, so both are close to unity and τ can be set zero only in the factor $(2 - i\tau^2)^{-\frac{1}{2}}$. When this is done the resulting approximation is

$$\sqrt{2}\,e^{-\frac{1}{4}i\pi}\,e^{-ik_0d}\int_{-\infty}^{\infty}\frac{\tau^2\,e^{-k_0d\tau^2}}{[\tau^2+i(1-\lambda)]\,[\tau^2+i(1-\lambda_0)]}\,d\tau. \qquad (4.78)$$

To handle (4.78), the non-exponential part of the integrand is written

$$-\frac{1}{\lambda-\lambda_0}\left[\frac{1-\lambda}{\tau^2+i(1-\lambda)}-\frac{1-\lambda_0}{\tau^2+i(1-\lambda_0)}\right],$$

and (3.51) employed. The approximation to $P_2(\lambda)$ is thus readily seen to be

$$e^{ik_0d\lambda_0}P_2(\lambda) = -\frac{1}{2\pi i}\frac{\sqrt{1-\lambda_0}\sqrt{1+\lambda}}{\lambda-\lambda_0}$$

$$-\frac{e^{-\frac{1}{4}i\pi}}{\pi\sqrt{2\pi}}\frac{\sqrt{1+\lambda_0}\sqrt{1+\lambda}}{\lambda-\lambda_0}e^{-ik_0d(1-\lambda_0)}$$

$$\times\{\sqrt{1-\lambda}\,F(\sqrt{k_0d(1-\lambda)})-\sqrt{1-\lambda_0}\,F(\sqrt{k_0d(1-\lambda_0)})\}. \quad (4.79)$$

The transmission cross-section is obtained from (4.75). The first term in (4.79) combined with the present approximation for $P_1(\lambda)$, which is just the first term in (4.72), yields the contribution $d\sin\alpha_0$ to σ_T, as before. To evaluate the remainder of (4.79) at $\lambda = \lambda_0$ requires the limit to be taken. With

$$\varepsilon = \sqrt{k_0d(1-\lambda_0)},$$

this limit is

$$\frac{e^{-\frac{1}{4}i\pi}}{2\pi\sqrt{2\pi}}\frac{\sqrt{k_0d}}{\varepsilon}\left(2-\frac{\varepsilon^2}{k_0d}\right)[F(\varepsilon)-\varepsilon+2i\varepsilon^2F(\varepsilon)]\,e^{-i\varepsilon^2}. \quad (4.80)$$

The correction to the geometrical cross-section $d\sin\alpha_0$ for σ_T is therefore the real part of $-2\pi/k_0$ times (4.80).

If $\varepsilon \ll 1$, (4.80) gives

$$\frac{e^{-\frac{1}{4}i\pi}}{\pi\sqrt{2\pi}}\frac{\sqrt{k_0d}}{\varepsilon}[\tfrac{1}{2}\sqrt{\pi}\,e^{-\frac{1}{4}i\pi}-2\varepsilon+\sqrt{\pi}\,e^{\frac{1}{4}i\pi}\varepsilon^2+O(\varepsilon^3)].$$

where, for consistency, only the leading explicit power of d is retained; and the contribution to the transmission cross-section is therefore

$$2\sqrt{\frac{d}{\pi k_0}}-\sqrt{\frac{2d}{k_0}}\,\varepsilon+O(\varepsilon^2). \quad (4.81)$$

To order ε, the second term in (4.81) cancels the other contribution, $d\sin\alpha_0$. Hence ,when $\alpha_0 \ll 1/\sqrt{k_0d}$,

$$\sigma_T = 2\sqrt{\frac{d}{\pi k_0}}+O(\varepsilon^2). \quad (4.82)$$

At grazing incidence $\varepsilon = 0$, and the leading term in the transmission cross-section is then proportional to the square root of the slit width.

4.2.4. The Narrow Slit

When a monochromatic wave is diffracted through an aperture whose linear dimensions are very much less than the wavelength the findings of theories owing some allegiance to geometrical optics do not constitute even a zero order approximation to the solution. The problem can be illustrated by considering a slit in a perfectly conducting plane, as in § 4.2.3, except that now $k_0 d \ll 1$. The isolated half-plane solution is no longer a useful signpost, and so the dual integral equations as given by the plane wave spectrum representation are used without preliminary transformation. To treat the simplest case we revert to the notation of § 4.1.3, taking the slit to occupy $y = 0$, $|x| < a$; and the plane wave to be E-polarized and normally incident. With the *total* field in $y < 0$ represented by (2.72), (2.73) with the lower sign, the boundary conditions at $y = 0$ give

$$\int_{-\infty}^{\infty} \frac{P(\lambda)}{\sqrt{1-\lambda^2}} e^{-ik_0 x\lambda}\, d\lambda = 0, \quad \text{for} \quad |x| > a, \tag{4.83}$$

$$\int_{-\infty}^{\infty} P(\lambda)\, e^{-ik_0 x\lambda}\, d\lambda = 1, \quad \text{for} \quad |x| < a. \tag{4.84}$$

From the symmetry of the problem $P(\lambda)$ is an even function of λ. If

$$\frac{P(\lambda)}{\sqrt{1-\lambda^2}} = \sum_{m=0}^{\infty} a_m \frac{J_{2m+1}(k_0 a\lambda)}{\lambda} \tag{4.85}$$

is chosen as a solution of (4.83), which it clearly is by virtue of the asymptotic behaviour of the Bessel functions as $\lambda \to \pm\infty$, and their freedom from singularities in the complex λ-plane, then substitution into (4.84) shows that the coefficients a_m are required to satisfy

$$\sum_{m=0}^{\infty} a_m \int_0^{\infty} \frac{\sqrt{1-\lambda^2}}{\lambda} J_{2m+1}(k_0 a\lambda) \cos(k_0 x\lambda)\, d\lambda = \frac{1}{2}, \quad \text{for} \quad |x| < a. \tag{4.86}$$

When $k_0a \ll 1$, the major contribution to the integrals in (4.86) comes from large values of λ, and the integrals are

$$i\int_0^\infty J_{2m+1}(k_0a\lambda)\cos(k_0x\lambda)\,d\lambda + O(k_0a)$$

$$= \frac{i\cos[(2m+1)\sin^{-1}(x/a)]}{k_0a\sqrt{1-x^2/a^2}} + O(k_0a),$$

the final integration being effected by use of the Bessel function representation quoted in (7.32).

Thus, to the lowest order approximation, (4.86) gives

$$a_0 = \frac{k_0a}{2i}, \quad a_m = 0 \quad \text{for} \quad m = 1, 2, 3\ldots \tag{4.87}$$

and correspondingly

$$\frac{P(\lambda)}{\sqrt{1-\lambda^2}} = \frac{k_0a}{2i}\,\frac{J_1(k_0a\lambda)}{\lambda}. \tag{4.88}$$

To this order the cross-section theorem gives zero for σ_T. But the time-averaged power transmitted through the slit per unit length can be evaluated from (3.4); division by $\frac{1}{2}Y_0$ gives

$$\sigma_T = \tfrac{1}{2}\pi k_0a^2\int_0^\pi \tan^2\alpha J_1^2(k_0a\cos\alpha)\,d\alpha,$$

and with $k_0a \ll 1$ the Bessel function can be replaced by $\frac{1}{2}k_0a\cos\alpha$ so that

$$\sigma_T = \frac{\pi^2}{16}k_0^3a^4. \tag{4.89}$$

4.2.5. Line-source

To conclude this chapter the case is considered of a line-source parallel to the edge of a perfectly conducting half-plane. The method of solution is simply to represent the source as a plane wave spectrum, and to apply the known solutions for plane wave incidence.

Let the faces of the half-plane be $\theta = 0$, $\theta = 2\pi$, as before, and let the line-source be specified by $\mathbf{H}^i = (0, 0, H_z^i)$, with

$$H_z^i = \sqrt{\frac{\pi}{2}}\,e^{-\frac{1}{4}i\pi}H_0^{(2)}(k_0R) \sim \frac{e^{-ik_0R}}{\sqrt{k_0R}}, \tag{4.90}$$

where R is distance from the location of the source at r_0, θ_0 $(0 < \theta_0 < \pi)$. Then, from (2.98), the incident field is written

$$H_z^i = \frac{e^{-\frac{1}{4}i\pi}}{\sqrt{2\pi}} \int_C e^{-ik_0 r_0 \cos(\theta_0-\alpha)} e^{ik_0 r \cos(\theta-\alpha)} \, d\alpha, \quad \text{for} \quad y \leqq r_0 \sin\theta_0, \tag{4.91}$$

where two points are worth noting: first, that the leading exponential in the integrand ensures that all the waves of the spectrum are in phase at r_0, θ_0; secondly, that the representation chosen is that valid throughout the half-space below the source, so that each plane wave of the spectrum is incident on the half-plane from above, in conformity with the analysis of § 4.2.2.

It is now evident that the solution can be obtained from that for an incident plane wave

$$\mathbf{H}^{pi} = (0, 0, 1)\, e^{ik_0 r \cos(\theta-\alpha)} \tag{4.92}$$

on multiplying by the factor

$$\frac{e^{-\frac{1}{4}i\pi}}{\sqrt{2\pi}} e^{-ik_0 r_0 \cos(\theta_0-\alpha)} \tag{4.93}$$

and integrating with respect to α. Admittedly, in § 4.2.2 no thought was given to the angle of incidence, there denoted by α_0, being other than real; but the solution (4.57) is an analytic function of α_0 without singularities, and is thus equally applicable to complex values. In fact, it is more convenient to use, rather than the form (4.57) itself, the corresponding angular spectrum representation. This can be written as the sum of the geometrical optics field and the diffracted field (cf. (4.12))

$$H_z^p = H_z^{pg} + H_z^{pd}, \tag{4.94}$$

where

$$H_z^{pd} = \frac{i}{\pi} \int_{S(0)} \frac{\cos\frac{\alpha}{2}\cos\left(\frac{\theta+\beta}{2}\right)}{\cos\alpha + \cos(\theta+\beta)} e^{-ik_0 r \cos\beta} \, d\beta, \tag{4.95}$$

using for H-polarization the angular spectrum function corresponding to (4.55), and with changes of notation consequent on α now being the angle of incidence.

If α is complex, with real part between 0 and π, the regions in which the different expressions for H_z^{pg} hold are determined by where the paths $S(\pi - \theta)$ and $S(\theta - \pi)$ in the complex α-plane lie relative to α. In multiplying (4.94) by (4.93), and integrating over α, it is therefore convenient to exercise the liberty to take the initial α path of integration as $S(\theta_0)$. This leads to the solution

$$H_z = H_z^g + H_z^d, \tag{4.96}$$

where

$$H_z^g = \begin{cases} \sqrt{\dfrac{\pi}{2}}\, e^{-\frac{1}{4}i\pi}\,[H_0^{(2)}(k_0R) + H_0^{(2)}(k_0S)], & \text{for } 0 < \theta < \pi - \theta_0, \\ \sqrt{\dfrac{\pi}{2}}\, e^{-\frac{1}{4}i\pi}\, H_0^{(2)}(k_0R), & \text{for } \pi - \theta_0 < \theta < \pi + \theta_0, \\ 0, & \text{for } \pi + \theta_0 < \theta < 2\pi, \end{cases} \tag{4.97}$$

S being distance from the image point r_0, $2\pi - \theta_0$, and

$$H_z^d = \frac{e^{\frac{1}{4}i\pi}}{\pi\sqrt{2\pi}} \int\limits_{S(0)}\int\limits_{S(0)} \frac{\cos\left(\dfrac{\alpha+\theta_0}{2}\right)\cos\left(\dfrac{\beta+\theta}{2}\right)}{\cos(\alpha+\theta_0) + \cos(\beta+\theta)} \times e^{-ik_0(r_0\cos\alpha + r\cos\beta)}\, d\alpha\, d\beta. \tag{4.98}$$

The double integral (4.98) can be simplified. First, write

$$H_z^d = I(\theta_0) + I(-\theta_0), \tag{4.99}$$

where

$$I(\theta_0) = \frac{e^{\frac{1}{4}i\pi}}{4\pi\sqrt{2\pi}} \int\limits_{S(0)}\int\limits_{S(0)} \frac{e^{-ik_0(r_0\cos\alpha + r\cos\beta)}}{\cos\left(\dfrac{\alpha+\beta+\theta_0+\theta}{2}\right)}\, d\alpha\, d\beta. \tag{4.100}$$

Then multiply both numerator and denominator of the integrand in (4.100) by $4\cos[\frac{1}{2}(\alpha - \beta + \theta_0 + \theta)]$, and discard that part of the resulting integrand that is an odd function of β. This gives

$$I(\theta_0) = \frac{e^{\frac{1}{4}i\pi}}{8\pi\sqrt{2\pi}} \int\limits_{S(0)}\int\limits_{S(0)} \left\{ \frac{1}{\cos\left(\dfrac{\alpha+\theta_0+\theta}{2}\right) - \sin\dfrac{\beta}{2}} + \frac{1}{\cos\left(\dfrac{\alpha+\theta_0+\theta}{2}\right) + \sin\dfrac{\beta}{2}} \right\} \cos\frac{\beta}{2}\, e^{-ik_0(r_0\cos\alpha + r\cos\beta)}\, d\alpha\, d\beta. \tag{4.101}$$

In the second term of the integrand in (4.101) change α to $-\alpha$; then

$$I(\theta_0) = \frac{e^{\frac{1}{4}i\pi}}{2\pi\sqrt{2\pi}} \int\limits_{S(0)} \int\limits_{S(0)}$$

$$\times \frac{\cos\left(\dfrac{\theta+\theta_0}{2}\right)\cos\dfrac{\alpha}{2}\cos\dfrac{\beta}{2}\, e^{-ik_0(r_0\cos\alpha + r\cos\beta)}}{\cos\alpha + \cos\beta - 2 - 4\sin\left(\dfrac{\theta_0+\theta}{2}\right)\sin\dfrac{\alpha}{2}\sin\dfrac{\beta}{2} + 2\cos^2\left(\dfrac{\theta_0+\theta}{2}\right)}$$

$$\times\, d\alpha\, d\beta. \tag{4.102}$$

The change of variables to ξ, η through

$$\xi = \sqrt{2}\, e^{-\frac{1}{4}i\pi} \sin\frac{\alpha}{2}, \quad \eta = \sqrt{2}\, e^{-\frac{1}{4}i\pi} \sin\frac{\beta}{2} \tag{4.103}$$

now gives

$$I(\theta_0) = -\frac{e^{\frac{1}{4}i\pi}}{\pi\sqrt{2\pi}}\, e^{-ik_0(r_0+r)} \cos\left(\frac{\theta_0+\theta}{2}\right)$$

$$\times \int\limits_{-\infty}^{\infty}\int\limits_{-\infty}^{\infty} \frac{e^{-k_0(r_0\xi^2+r\eta^2)}}{\xi^2+\eta^2+2\sin\left(\dfrac{\theta_0+\theta}{2}\right)\xi\eta + 2i\cos^2\left(\dfrac{\theta_0+\theta}{2}\right)}\, d\xi\, d\eta; \tag{4.104}$$

and the further change to ϱ, φ through

$$\xi = \sqrt{R_1/r_0}\,\varrho\cos\varphi, \quad \eta = \sqrt{R_1/r}\,\varrho\sin\varphi, \tag{4.105}$$

where $R_1 = r_0 + r$, yields

$$I(\theta_0) = -\frac{e^{\frac{1}{4}i\pi}}{\pi\sqrt{2\pi}}\, e^{-ik_0R_1} \cos\left(\frac{\theta_0+\theta}{2}\right) \int\limits_0^{\infty} \varrho K(\varrho)\, e^{-k_0R_1\varrho^2}\, d\varrho, \tag{4.106}$$

with

$$K(\varrho) = \int\limits_0^{2\pi} \left\{\varrho^2\left[\sqrt{\frac{r}{r_0}}\cos^2\varphi + \sqrt{\frac{r_0}{r}}\sin^2\varphi\right.\right.$$

$$\left.\left. + 2\sin\left(\frac{\theta_0+\theta}{2}\right)\sin\varphi\cos\varphi\right] + 2i\frac{\sqrt{r_0r}}{R_1}\cos^2\left(\frac{\theta_0+\theta}{2}\right)\right\}^{-1} d\varphi. \tag{4.107}$$

Evidently $K(\varrho)$ can be evaluated by putting $\zeta = \exp(i\varphi)$, so that

$$K(\varrho) = -4i \int_{\substack{\text{unit}\\ \text{circle}}} \frac{\zeta\, d\zeta}{A\zeta^4 + 2B\zeta^2 + C}, \tag{4.108}$$

where

$$A = \sqrt{\frac{r}{r_0}} - \sqrt{\frac{r_0}{r}} - 2i \sin\left(\frac{\theta_0 + \theta}{2}\right),$$

$$B = \frac{R_1}{\sqrt{rr_0}}\varrho^2 + 4i\frac{\sqrt{rr_0}}{R_1}\cos^2\left(\frac{\theta_0 + \theta}{2}\right),$$

$$C = \sqrt{\frac{r}{r_0}} - \sqrt{\frac{r_0}{r}} + 2i \sin\left(\frac{\theta_0 + \theta}{2}\right).$$

It is easily confirmed that two of the poles of the integrand in (4.108) are inside the unit circle, and that two are outside, with the result

$$K(\varrho) = 2\pi \left|\sec\left(\frac{\theta + \theta_0}{2}\right)\right| \Big/ \sqrt{\varrho^4 + 2i\varrho^2 - \frac{4rr_0}{R_1^2}\cos^2\left(\frac{\theta_0 + \theta}{2}\right)}, \tag{4.109}$$

where the square root is in the first quadrant. The substitution of (4.109) into (4.106) gives

$$I(\theta_0) = \mp \sqrt{\frac{2}{\pi}}\, e^{\frac{1}{4}i\pi}\, e^{-ik_0R_1}$$

$$\times \int_0^\infty \frac{\varrho\, e^{-k_0R_1\varrho^2}}{\sqrt{\left(\varrho^2 + i\dfrac{R_1 - S}{R_1}\right)\left(\varrho^2 + i\dfrac{R_1 + S}{R_1}\right)}}\, d\varrho, \tag{4.110}$$

with upper/lower sign for $\cos\left[\frac{1}{2}(\theta_0 + \theta)\right] \gtrless 0$.

By adopting the form

$$I(\theta_0) = \mp \sqrt{\frac{2}{\pi}}\, e^{\frac{1}{4}i\pi}\, e^{-ik_0S} \int_{\sqrt{k_0(R_1 - S)}}^{\infty} \frac{e^{-i\lambda^2}}{\sqrt{\lambda^2 + 2k_0S}}\, d\lambda, \tag{4.111}$$

with the analogous expression for $I(-\theta_0)$, and noting that

$$H_0^{(2)}(k_0R) = \frac{2i}{\pi}\, e^{-ik_0R} \int_{-\infty}^{\infty} \frac{e^{-i\lambda^2}}{\sqrt{\lambda^2 + 2k_0R}}\, d\lambda, \tag{4.112}$$

the total field can be written

$$H_z = \sqrt{\frac{2}{\pi}}\, e^{\frac{1}{4}i\pi} \left\{ e^{-ik_0R} \int_{-p}^{\infty} \frac{e^{-i\lambda^2}}{\sqrt{\lambda^2 + 2k_0R}}\, d\lambda \right.$$

$$\left. + \, e^{-ik_0S} \int_{-q}^{\infty} \frac{e^{-i\lambda^2}}{\sqrt{\lambda^2 + 2k_0S}}\, d\lambda \right\}, \qquad (4.113)$$

where

$$p = 2\sqrt{\frac{k_0r_0r}{r_0 + r + R}} \cos\left(\frac{\theta_0 - \theta}{2}\right),$$

$$q = 2\sqrt{\frac{k_0r_0r}{r_0 + r + S}} \cos\left(\frac{\theta_0 + \theta}{2}\right). \qquad (4.114)$$

The form (4.113) is very similar to (4.57), to which it clearly reverts when, after multiplication by $\sqrt{k_0r_0}\exp(ik_0r_0)$, r_0 is made to tend to infinity. It may also be noted that a Fresnel integral approximation for the diffracted field is valid provided only that $k_0R_1 \gg 1$; for example, in (4.111) the non-exponential part of the integrand can then be replaced by its value at the lower limit, to give the approximation

$$I(\theta_0) = \mp \sqrt{\frac{2}{\pi}}\, e^{\frac{1}{4}i\pi} \frac{e^{-ik_0R_1}}{\sqrt{k_0(R_1 + S)}} F\{\sqrt{k_0(R_1 - S)}\}. \qquad (4.115)$$

As a final comment it is worth drawing attention to the fact that the spectrum function $P(\cos\alpha)$ resulting from a plane wave incident at angle α_0, (4.55) for example, is symmetric in α_0 and α. This is a manifestation of the reciprocity inherent in Maxwell's equations, which in the present context signifies that the solution for a line-source is unaffected if the location of the source is interchanged with that of the point of observation ($r_0 \leftrightarrow r$, $\theta_0 \leftrightarrow \theta$).

CHAPTER V

PROPAGATION OVER A UNIFORM PLANE SURFACE

5.1. RADIO PROPAGATION OVER A HOMOGENEOUS EARTH

5.1.1. Reflection Coefficients for Plane Wave Incidence

When radio waves transmitted from one point on or near the surface of the earth are received at another, there are, of course, many factors affecting the strength of the received signal. One question of a rather general nature is the extent to which the electrical properties of the ground, represented by its permittivity and conductivity, control the behaviour of the field. A comparatively simple idealized problem, whose solution goes some way to answering this question, at least when the distances involved are not so great that earth curvature has to be taken into account, is that in which the source is situated in a vacuum above the plane boundary of a semi-infinite, homogeneous medium.

The main interest of the problem is certainly covered by considering distances from the source greater than a wavelength or so, and it might be thought that the answer would be trivially obtained from a straightforward "ray" treatment, in which allowance is made for the earth's reflection coefficient. That this is by no means always the case is mainly because, for a certain polarization, the reflection coefficient is very sensitive to the angle of incidence when the incidence is near grazing, and, since the distance between transmitter and receiver is commonly much greater than the sum of their heights above the earth's surface, the effective angle of incidence is likely to be small. It is instructive, therefore, before proceeding with the full analysis, to recall briefly the properties of the Fresnel reflection coefficients.

As explained in § 1.2, for a given angular frequency ω, the permittivity ε' and conductivity σ of the medium appear in the analysis only in the combination $\varepsilon' - i\sigma/\omega$. It is convenient now to denote this parameter, the complex permittivity, by the symbol ε. The permeability of the medium is assumed not to differ from the vacuum permeability μ_0, and the symbol μ is reserved for the complex refractive index (called μ_c in § 2.1.3), namely

$$\mu = \sqrt{\varepsilon/\varepsilon_0}\,. \tag{5.1}$$

It is assumed, as would be the case for any earth constituent, that ε' and σ are positive. Then ε lies somewhere in the fourth quadrant of the complex plane, and the convention for the radical in (5.1) is that it is the branch that is positive when ε is real; thus μ has argument between 0 and $-\frac{1}{4}\pi$. The characteristic impedance of the medium $Z = \sqrt{\mu_0/\varepsilon}$ and admittance $Y = 1/Z$ are, in general, complex; they are related to the corresponding vacuum quantities through

$$Z = Z_0/\mu, \quad Y = \mu Y_0\,. \tag{5.2}$$

The propagation coefficient $k = \omega\sqrt{\varepsilon\mu_0}$ is likewise complex; evidently

$$k = k_0\mu\,. \tag{5.3}$$

Suppose now that a homogeneous plane wave in the vacuum falls on the plane face of the semi-infinite medium at angle of incidence α (normal incidence being $\alpha = \frac{1}{2}\pi$). There are, in general, reflected and transmitted waves, and the reflection and transmission coefficients depend on the polarization. For an E-polarized incident wave the calculation is virtually that given in § 2.1.5; the reflection coefficient is

$$\varrho_E = \frac{Y_0 \sin\alpha - Y\sin\beta}{Y_0 \sin\alpha + Y\sin\beta}, \tag{5.4}$$

where

$$k_0 \cos\alpha = k\cos\beta\,. \tag{5.5}$$

For an H-polarized wave the reflection coefficient is

$$\varrho_H = \frac{Z_0 \sin\alpha - Z\sin\beta}{Z_0 \sin\alpha + Z\sin\beta}\,. \tag{5.6}$$

In the former case $\mathbf{E}$ is perpendicular to the plane of incidence, in the latter case parallel to it. In the context of propagation over the earth these cases are sometimes referred to as horizontal and vertical polarization respectively. The formulae as written are quite general; but if the particular relations (5.2), (5.3) appropriate to vacuum permeability are used, they appear as

$$\varrho_E = \frac{\sin\alpha - \mu\sin\beta}{\sin\alpha + \mu\sin\beta}, \tag{5.7}$$

$$\varrho_H = \frac{\mu\sin\alpha - \sin\beta}{\mu\sin\alpha + \sin\beta}, \tag{5.8}$$

where

$$\cos\alpha = \mu\cos\beta. \tag{5.9}$$

When a source is represented as a plane wave spectrum it is necessary to understand the behaviour of (5.7) and (5.8) as functions of α. Since

$$\sin\beta = \sqrt{1 - \frac{\cos^2\alpha}{\mu^2}}, \tag{5.10}$$

both these functions have branch-points at

$$\cos\alpha = \pm\mu. \tag{5.11}$$

As regards the possession of poles or zeros they are, however, quite different in character.

Consider, first, (5.7). If it has any poles or zeros they must be at values of α that satisfy

$$\sin^2\alpha = \mu^2\sin^2\beta; \tag{5.12}$$

but since, from (5.10),

$$\mu^2\sin^2\beta = \mu^2 - \cos^2\alpha,$$

the satisfaction of (5.12) is impossible, save for the trivial case $\mu = 1$. Thus ϱ_E has no poles. Furthermore, for the earth at radio frequencies it can be taken that $|\mu| \gg 1$. In this case, for all real values of α, $\sin\beta$ is close to unity, from (5.10), and ϱ_E is close to -1. The reflection coefficient is in no way sensitive to the angle of incidence.

It is another matter, however, with (5.8). Here possible poles or zeros occur at values of α that satisfy

$$\mu^2\sin^2\alpha = \sin^2\beta, \tag{5.13}$$

which gives

$$\sin\alpha = \pm\frac{1}{\sqrt{\mu^2+1}}. \tag{5.14}$$

If the convention is adopted that the radical in (5.14) is positive when μ is real, a glance back at (5.10) and (5.8) is sufficient to establish that ϱ_H is zero when

$$\sin\alpha = \frac{1}{\sqrt{\mu^2+1}}, \tag{5.15}$$

and has poles at

$$\sin\alpha = -\frac{1}{\sqrt{\mu^2+1}}. \tag{5.16}$$

If μ is real, (5.15) shows that there is a real angle of incidence for which there is no reflected wave. This is the familiar ***Brewster angle***, and is easily seen to be that angle for which the directions of reflected wave (were there one) and transmitted wave are orthogonal. For $\mu \gg 1$, the transmitted wave travels almost at right angles to the interface, and Brewster incidence is near grazing.

When μ is not real, the counterpart to the Brewster angle is complex. If it is denoted by α_B, with $0 < \operatorname{Re}\alpha_B < \frac{1}{2}\pi$, then (5.16). states that ϱ_H has poles at the values of α for which

$$\sin\alpha = -\sin\alpha_B. \tag{5.17}$$

Since μ has argument between 0 and $-\frac{1}{4}\pi$, it appears that these poles lie somewhere in the regions of the complex α-plane shown shaded in Fig. 5.1; it should be noted, in particular, that these regions are outside the strip $0 < \operatorname{Re}\alpha < \pi$, and that the tangents to their curved boundaries at $\alpha = 0$ and $\alpha = \pi$ make angles $\frac{1}{4}\pi$ with the axes. A consequence of $|\mu| \gg 1$ is that $|\alpha_B| \ll 1$, and there is a pole close to $\alpha = 0$. The effect of this is to make ϱ_H vary rapidly

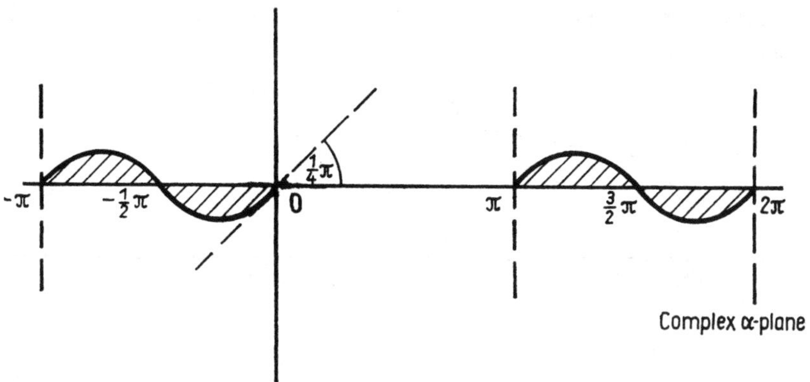

FIG. 5.1.

with α when α becomes small; indeed, it is apparent from (5.8) that whereas ϱ_H is -1 at $\alpha = 0$, it swings over to a value close to $+1$ as soon as α increases much beyond $1/|\mu|$. It is just this behaviour of the reflection coefficient which gives the problem of a vertically polarized source above the earth's surface its special interest.

5.1.2. Solution for a Localized Source: *E*-polarization

The problem now treated explicitly is that in which a line-source is situated above the semi-infinite homogeneous medium. However, the solution is put into a form that applies also to the case of a point-source.

Consider the case in which the source is *E*-polarized, and specified by

$$E_z^i = \sqrt{\tfrac{1}{2}\pi}\, e^{-\frac{1}{4}i\pi}\, H_0^{(2)}(k_0 R) \sim \frac{e^{-ik_0 R}}{\sqrt{k_0 R}}, \tag{5.18}$$

where R denotes distance from the location of the source at r_0, θ_0 $(0 < \theta_0 < \pi)$. Then, just as in § 4.2.5, the plane wave spectrum representation of (5.18) is introduced; namely, the right-hand side of (4.91). Each plane wave of the spectrum gives rise to a reflected wave, with reflection coefficient (5.7). Hence, in $y > 0$, there must be superposed on (5.18) the reflected field

$$E_z^r = \frac{e^{-\frac{1}{4}i\pi}}{\sqrt{2\pi}} \int_C \varrho_E^{(\alpha)}\, e^{ik_0 S \cos(\psi+\alpha)}\, d\alpha, \tag{5.19}$$

where S, $\psi (0 < \psi < \frac{1}{2}\pi$, say) are polar coordinates of the point of observation measured from the source's image point r_0, $-\theta_0$ (see Fig. 5.2).

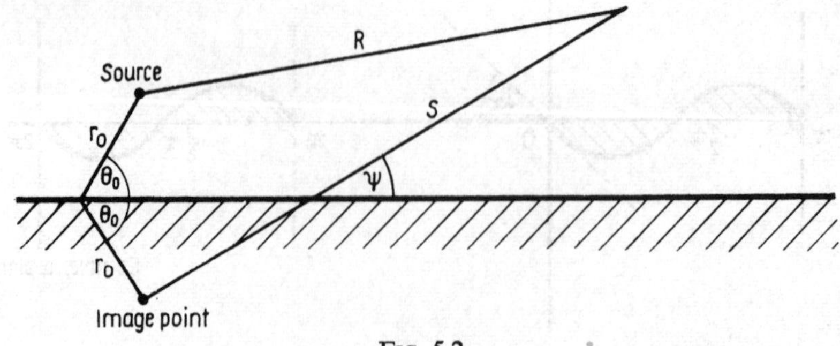

FIG. 5.2.

From (5.7),

$$\varrho_E = -1 + \frac{2 \sin \alpha}{\sin \alpha + \mu \sin \beta}, \tag{5.20}$$

so that the total field in $y > 0$ can be written

$$E_z = \sqrt{\tfrac{1}{2}\pi}\, e^{-\frac{1}{4}i\pi} [H_0^{(2)}(k_0 R) - H_0^{(2)}(k_0 S)] + E_{1z}, \tag{5.21}$$

where

$$E_{1z} = \sqrt{\frac{2}{\pi}}\, e^{-\frac{1}{4}i\pi} \int\limits_C \frac{\sin \alpha}{\sin \alpha + \mu \sin \beta} e^{ik_0 S \cos (\psi + \alpha)}\, d\alpha. \tag{5.22}$$

In (5.21) the field is expressed as the superposition of a field E_{1z} on that which would exist were the medium a perfect conductor.

The natural step to take next, particularly with the condition $k_0 S \gg 1$ in mind, is to distort the path of integration C in (5.22) into the path of steepest descents $S(\pi - \psi)$. But in doing this attention must be paid to the branch-points. Recalling that the argument of μ is between 0 and $-\frac{1}{4}\pi$, it is seen that with $|\mu| \gg 1$ the branch-points are located somewhat as indicated in Fig. 5.3. Apart, then, from the exceptional case when ψ is close to $\frac{1}{2}\pi$, one of the branch-points $\cos \alpha = -\mu$ does indeed lie between C and $S(\pi - \psi)$. Strictly, therefore, an integral round the corresponding branch-cut should be included in the solution. However, it can be argued that this branch-cut contribution is negligible in the common case when the conductivity of the medium is sufficient to

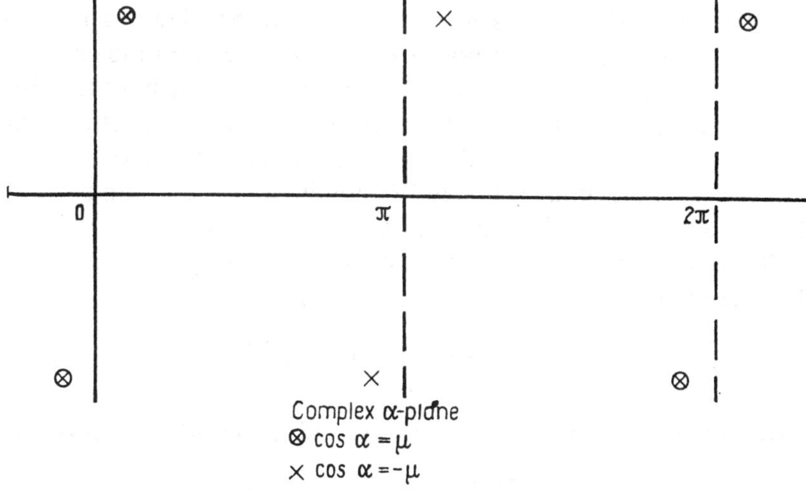

FIG. 5.3.

give μ an appreciable imaginary part; for its order of magnitude is evidently determined by the factor $\exp[ik_0 S\cos(\psi + \alpha)]$ evaluated at the branch-point, which factor is, for example, $\exp(-ik_0 S\mu)$ when $\psi = 0$. A closer look at the branch-cut contribution is postponed until a little later, and for the moment it is neglected without more ado. The leading term in the asymptotic approximation to (5.22) when $k_0 S \gg 1$ then appears as

$$E_{1z} = \frac{2\sin\psi}{\sin\psi + \sqrt{\mu^2 - \cos^2\psi}} \frac{e^{-ik_0 S}}{\sqrt{k_0 S}}. \tag{5.23}$$

It is convenient to express the total field in the form

$$E_z = \Delta E_z^i. \tag{5.24}$$

Δ is the factor by which the vacuum field of the isolated source must be multiplied in order to obtain the actual field. With allowance, if necessary, for the angular dependence of the radiation field of the source, the expressions to be given for Δ when $k_0 R \gg 1$ are equally applicable to the case of a point-source.

Evidently (5.21) and (5.23) give, for $k_0 R \gg 1$,

$$\Delta = 1 + \varrho_E(\psi)\, e^{-ik_0(S-R)}. \tag{5.25}$$

This is nothing other than the result of a simple-minded "ray" theory, and is, indeed, the main formula for E-polarization.

If both source and point of observation are at ground level, distance d apart, $S = R = d$ and $\psi = 0$. Then (5.25) is zero, and a better approximation is required. This can be obtained by proceeding to the next inverse power of $k_0 d$ in the asymptotic development of (5.22). On the other hand, when $\psi = 0$ it happens that (5.22) can be expressed exactly in terms of a Hankel function. The latter form proves useful subsequently, and is now derived.

Multiply both numerator and denominator of the integrand in (5.22) by $\sin\alpha - \mu\sin\beta$, and discard that part which is odd in α and therefore integrates to zero over the path $S(\pi)$. Thus, with the help of (5.10),

$$E_z = E_{1z} = -\sqrt{\frac{2}{\pi}}\, e^{-\frac{1}{4}i\pi} \frac{1}{\mu^2 - 1} \int_{S(\pi)} \sin^2\alpha\, e^{ik_0 d\cos\alpha}\, d\alpha. \tag{5.26}$$

From the corresponding integral form of $H_0^{(2)}(k_0 d)$ it follows that

$$E_z = -\frac{\sqrt{2\pi}\, e^{-\frac{1}{4}i\pi}}{\mu^2 - 1} [H_0^{(2)}(k_0 d) + H_0^{(2)\prime\prime}(k_0 d)], \tag{5.27}$$

a dash denoting differentiation with respect to the argument; and then, from Bessel's equation, that

$$E_z = \frac{\sqrt{2\pi}\, e^{-\frac{1}{4}i\pi}}{\mu^2 - 1} \frac{H_0^{(2)\prime}(k_0 d)}{k_0 d}. \tag{5.28}$$

The first term of the asymptotic development for $k_0 d \gg 1$ is therefore

$$E_z = -\frac{2i}{\mu^2 - 1} \frac{e^{-ik_0 d}}{(k_0 d)^{3/2}}, \tag{5.29}$$

and correspondingly

$$\Delta = -\frac{2i}{\mu^2 - 1} \frac{1}{k_0 d}. \tag{5.30}$$

Under certain conditions the forms (5.30) and (5.25) can be reconciled by means of the so-called *height-gain* function. For a plane wave near glancing incidence it is easy to confirm that the field in the vicinity of the surface is given in terms of the field at the surface, for the same value of x, by the formula

$$E_z(y) = \left[1 + ik_0 \sqrt{\mu^2 - 1}\, y\right] E_z(0), \quad \text{for} \quad y > 0. \tag{5.31}$$

Since (5.31) is independent of the angle of incidence it may be expected to apply also to the line-source field. This is so. The detailed derivation is given later for the analogous result for H-polarization, and here it is merely noted that (5.31) is in effect the same as the often quoted "impedance" boundary condition

$$\frac{\partial E_z}{\partial y} = ik_0 \mu E_z, \quad \text{at} \quad y = 0, \tag{5.32}$$

valid, approximately, when $|\mu| \gg 1$. The factor

$$1 + ik_0 \sqrt{\mu^2 - 1}\, y \tag{5.33}$$

is called the height-gain function. With $|\mu| \gg 1$ it represents a rapid increase of field strength as the point of observation is raised from the surface. If $k_0 d$ is sufficiently large (5.31) remains applicable when $k_0 \sqrt{\mu^2 - 1}\, y \gg 1$, so that in these circumstances the value of Δ corresponding to (5.30) for an elevated receiver is

$$\Delta = \frac{2}{\sqrt{\mu^2 - 1}} \frac{y}{d}; \tag{5.34}$$

and this is just the approximation given by (5.25) when $S = R$ and $\sin \psi = y/d \ll 1$.

By reciprocity, the height-gain factor (5.33), with y_0 replacing y, applies equally to the elevation of the transmitter. With y_0 as well as y sufficiently large, (5.34) is modified to

$$\Delta = 2ik_0 y_0 y/d, \tag{5.35}$$

which is the "ray" theory result with reflection coefficient -1.

In conclusion, it is noted that, when $y_0 = y = 0$, a complete exact solution, on the lines of (5.28), can be obtained. This is of particular value in showing explicitly the branch-cut contribution. To get the solution it is only necessary to follow the derivation of (5.26) from (5.22), but with C retained as the path of integration. The price paid for keeping the branch-cut contribution is that the odd part of the integrand no longer integrates to zero, and so

$$E_z = -\sqrt{\frac{2}{\pi}}\, e^{-\frac{1}{4}i\pi} \frac{1}{\mu^2 - 1}$$
$$\times \int_C [\sin^2\alpha - \sin\alpha \sqrt{\mu^2 - \cos^2\alpha}]\, e^{ik_0 d \cos\alpha}\, d\alpha. \tag{5.36}$$

Now the substitution $\cos\alpha = \mu\cos\beta$ gives

$$\int \sin\alpha \sqrt{\mu^2 - \cos^2\alpha}\, e^{ik_0 d\cos\alpha}\, d\alpha = \mu^2 \int \sin^2\beta\, e^{ik_0 d\mu\cos\beta}\, d\beta, \tag{5.37}$$

where the paths of integration can be chosen to ensure convergence, and need not be particularized. But the integral on the right-hand side of (5.37) is just that in (5.26), with $k_0 d\mu$ replacing $k_0 d$. An exact expression for the total field, corresponding to the approximation (5.28), is therefore

$$E_z = \sqrt{2\pi}\, e^{-\frac{1}{4}i\pi} \frac{H_0^{(2)\prime}(k_0 d) - \mu H_0^{(2)\prime}(k_0 \mu d)}{(\mu^2 - 1)\, k_0 d}. \tag{5.38}$$

When $k_0 d \gg 1$ the leading term gives

$$\Delta = -2i \frac{1 - \sqrt{\mu}\, e^{-ik_0 d(\mu - 1)}}{(\mu^2 - 1)\, k_0 d}. \tag{5.39}$$

The difference between (5.39) and (5.30) is precisely the branch-cut contribution. As anticipated, it does, indeed, contain the factor $\exp(-ik_0\mu d)$, and is therefore likely to be negligible in practice. However, in the ideal case in which μ is real (and greater than unity), the branch-cut contribution is dominant by virtue of the factor $\sqrt{\mu}$. For an H-polarized line-source, considered next, no simple exact solution is available, but the present information is sufficient to demonstrate that the analogous branch-cut contribution is negligible provided only that $|\mu| \gg 1$.

5.1.3. Solution for a Localized Source: *H*-polarization

The problem now considered is the same as that of § 5.1.2, except that the line-source is H-polarized, being specified by

$$H_z^i = \sqrt{\tfrac{1}{2}\pi}\, e^{-\frac{1}{4}i\pi}\, H_0^{(2)}(k_0 R) \sim \frac{e^{-ik_0R}}{\sqrt{k_0 R}}. \tag{5.40}$$

The associated reflected field in $y > 0$ is

$$H_z^r = \frac{e^{-\frac{1}{4}i\pi}}{\sqrt{2\pi}} \int_C \varrho_H(\alpha)\, e^{ik_0 S \cos(\psi+\alpha)}\, d\alpha. \tag{5.41}$$

From (5.8),

$$\varrho_H = 1 - \frac{2\sin\beta}{\mu\sin\alpha + \sin\beta}, \tag{5.42}$$

so that the total field in $y > 0$ can be written

$$H_z = \sqrt{\tfrac{1}{2}\pi}\, e^{-\frac{1}{4}i\pi}[H_0^{(2)}(k_0R) + H_0^{(2)}(k_0S)] + H_{1z}, \tag{5.43}$$

where

$$H_{1z} = -\sqrt{\frac{2}{\pi}}\, e^{-\frac{1}{4}i\pi} \int_C \frac{\sin\beta}{\mu\sin\alpha + \sin\beta}\, e^{ik_0S\cos(\psi+\alpha)}\, d\alpha. \tag{5.44}$$

In (5.43) the field is expressed as the superposition of a field H_{1z} on that which would exist were the medium a perfect conductor.

To evaluate (5.44) for $k_0 S \gg 1$ the aim is to distort the path of integration into the steepest descents path $S(\pi - \psi)$. The integrand now has poles at $\sin\alpha = -\sin\alpha_B$, as well as branch-points at $\cos\alpha = \pm\mu$. The configuration in the complex α plane for $|\mu| \gg 1$ is sketched in Fig. 5.4.

Just as in the treatment of E-polarization, a branch-point at $\cos\alpha = -\mu$ lies between C and $S(\pi - \psi)$ for the values of ψ of most interest. Again, however, the contribution of the associated branch-cut integral is legitimately neglected. The grounds for this are, indeed, stronger in the present case; for not only is there equal force in the argument for ignoring the contribution when μ has an appreciable imaginary part, but there is also an alternative argument that applies even if μ is allowed to be real. The latter relies on the inequality $|\mu| \gg 1$, and can be based on the analysis given towards the end of § 5.1.2. The case $\psi = 0$ no longer yields an exact solution, such as (5.38), but a parallel treatment shows that the branch-cut contribution is down by a factor of the order $1/|\mu|^2$ on the previous result. That is to say, from (5.39), it is of order $\mu^{-7/2}/(k_0 d)^{3/2}$ when μ is real. This is negligible compared with the main field, now to be obtained, which turns out to be at least of order $\mu^2/(k_0 d)^{3/2}$.

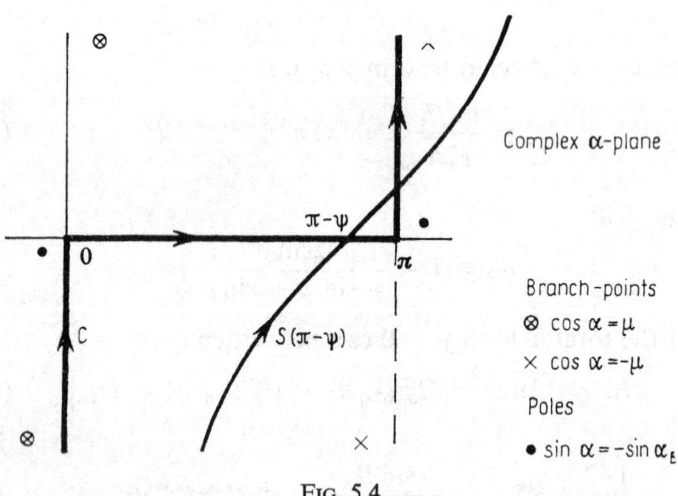

FIG. 5.4.

With the branch-points disposed of, what about the poles? The first thing to notice is that, for the problem in hand, they cannot lie between C and $S(\pi - \psi)$. This is clear from Fig. 5.1; for since $S(\pi - \psi)$ has least slope, unity, where it passes through $\pi - \psi$, it cannot enter the region to which the poles are confined. On the other hand, with $|\mu| \gg 1$, one of the poles is close to π, and hence close to the saddle-point when ψ is small. This must be taken into account in the integration by steepest descents, and the method of § 3.3.3 applied.

The procedure can therefore be detailed as follows. In (5.44), distort C to $S(\pi - \psi)$, ignoring the branch-cut integral, and then remove from under the integral sign at $\alpha = \pi - \psi$ the non-exponential part of the integrand, with the exception of the factor $\sec[\frac{1}{2}(\alpha - \alpha_B)]$ that contains the relevant pole. The resulting approximation for $k_0 S \gg 1$ is (cf. (3.71), (3.72))

$$H_{1z} = -2\sqrt{2}\, i\varrho_1(\psi) \sec\left(\frac{\psi - \alpha_B}{2}\right) e^{-ik_0 S} F\left[\sqrt{2k_0 S} \sin\left(\frac{\psi + \alpha_B}{2}\right)\right], \tag{5.45}$$

where

$$\varrho_1(\psi) = \frac{1}{\mu}\sqrt{1 - \frac{\cos^2\psi}{\mu^2}} \frac{\sin\psi + \sin\alpha_B}{\sin\psi + \frac{1}{\mu}\sqrt{1 - \frac{\cos^2\psi}{\mu^2}}}. \tag{5.46}$$

An algebraically simpler version of this result can be obtained by using the fact that ψ and $|\alpha_B|$ are small. Both $\sin\alpha_B$, given by (5.15), and $(1/\mu)\sqrt{1 - \cos^2\psi/\mu^2}$ can be approximated by $(1/\mu) \times \sqrt{1 - 1/\mu^2}$; so that (5.46) reads simply

$$\varrho_1(\psi) = \sin\alpha_B.$$

Also $\sin[\frac{1}{2}(\psi + \alpha_B)]$ can be replaced by $\frac{1}{2}(\sin\psi + \sin\alpha_B)$, and $\sec[\frac{1}{2}(\psi - \alpha_B)]$ by unity; for even when these approximations are not in themselves very accurate, they remain adequate when taken together. Thus

$$H_{1z} = -2\sqrt{2}\, i \sin\alpha_B\, e^{-ik_0 S} F\left[\sqrt{\tfrac{1}{2}k_0 S}\,(\sin\psi + \sin\alpha_B)\right]. \tag{5.47}$$

Summing up, for $k_0 R \gg 1$ the total field in $y > 0$ is given by

$$H_z = \Delta H_z^i, \tag{5.48}$$

where

$$\Delta = 1 + [1 - 4i\gamma_0 F(\gamma)]\, e^{-ik_0(S-R)}, \tag{5.49}$$

with

$$\gamma = \sqrt{\tfrac{1}{2}k_0 S}\,(\sin\psi + \sin\alpha_B), \tag{5.50}$$

$$\gamma_0 = \sqrt{\tfrac{1}{2}k_0 S}\,\sin\alpha_B. \tag{5.51}$$

These formulae are equally applicable to the case of a point-source.

5.1.4. Special Cases

It is convenient to begin by considering the case when both source and point of observation lie on the surface; $\psi = 0$ and $S = R = d$. Then (5.49) is

$$\Delta = 2[1 - 2i\gamma_0 F(\gamma_0)] \tag{5.52}$$

with $\gamma_0 = \sqrt{\frac{1}{2}k_0 d}\,\sin\alpha_B$.

In general, γ_0 is complex. It is real when the conductivity is zero; and, at the other extreme, it has argument effectively $\frac{1}{4}\pi$ when the complex permittivity is dominated by the conductivity, as, for example, with sea water at comparatively low frequencies. In this latter case (3.46) shows that (5.52) can be written in terms of an integral with real limits, thus

$$\Delta = 2\left[1 - |\gamma_0|\, e^{-|\gamma_0|^2}\left(i\sqrt{\pi} + 2\int_0^{|\gamma_0|} e^{\lambda^2}\, d\lambda\right)\right]. \tag{5.53}$$

The result is often quoted in this form, with $\gamma_0 \exp(-\frac{1}{4}i\pi)$ replacing $|\gamma_0|$, even when $\arg \gamma_0$ is less than $\frac{1}{4}\pi$.

Traditionally, $|\gamma_0|$ is called the "numerical distance", the point being that the nature of (5.52) is quite different according to whether $|\gamma_0|$ is small or large, and that $|\gamma_0|$ may be small even though $k_0 d \gg 1$.

When $|\gamma_0| \ll 1$,

$$\Delta = 2(1 - e^{\frac{1}{4}i\pi}\sqrt{\pi}\,\gamma_0) + O(\gamma_0^2). \tag{5.54}$$

The field is therefore not very different from what it would be were the medium a perfect conductor.

When $|\gamma_0| \gg 1$, the asymptotic expansion (3.56) can be used for the Fresnel integral. The first term alone would give the approximation zero for (5.52). The first two terms give

$$\Delta = -\frac{i}{\gamma_0^2} = -\frac{2i}{k_0 d \sin^2 \alpha_B}. \tag{5.55}$$

Revert now to a consideration of the general formula (5.49). When $|\gamma| \ll 1$, which of course implies $|\gamma_0| \ll 1$,

$$\Delta = 1 + (1 - 2e^{\frac{1}{4}i\pi}\gamma_0)\, e^{-ik_0(S-R)}, \tag{5.56}$$

again close to the result for a perfect conductor.

When $|\gamma| \gg 1$, use of the first term alone of the asymptotic expansion of the Fresnel integral gives

$$\Delta = 1 + \frac{\sin\psi - \sin\alpha_B}{\sin\psi + \sin\alpha_B}\, e^{-ik_0(S-R)}. \tag{5.57}$$

This is evidently what would be obtained from "ray" theory, with the appropriate approximate form of the reflection coefficient. As ψ approaches zero, the reflected wave more nearly cancels the incident, and the second term of the asymptotic expansion must be brought in. Then

$$\Delta = 1 + \left\{\frac{\sin\psi - \sin\alpha_B}{\sin\psi + \sin\alpha_B} - \frac{2i\sin\alpha_B}{(\sin\psi + \sin\alpha_B)^3}\frac{1}{k_0 S}\right\} e^{-ik_0(S-R)}, \tag{5.58}$$

from which (5.55) is recovered when $\psi = 0$.

To get some idea of when (5.57) alone is adequate, consider the case $y_0 = 0$, so that $S = R$. Then with $\sin\psi$ written as y/S, and assumed small compared to $|\sin\alpha_B|$ (for if not, (5.57) is certainly satisfactory), the extra term in (5.58) is negligible if and only if

$$k_0 y |\sin\alpha_B| \gg 1. \tag{5.59}$$

When $y_0 \neq 0$ the corresponding condition

$$k_0(y_0 + y)|\sin\alpha_B| \gg 1 \tag{5.60}$$

is certainly sufficient, but may demand more than is strictly necessary, since with $y_0 = y$, for example, the condition ensures that it is the difference between $\exp[-ik_0(S - R)]$ and unity that dominates (5.58).

It is also possible, as for E-polarization, to derive the field for limited elevations of transmitter and receiver from the application of a height-gain function to (5.55). This is now established directly from the plane wave spectrum representation of the field.

With y_0 and y introduced explicitly, the integral representation of the incident field, taken over the steepest descents path, can be written

$$H_z^i = \frac{e^{-\frac{1}{4}i\pi}}{\sqrt{2\pi}} \int_{S(0)} e^{-ik_0 d\cos\alpha}\, e^{-ik_0|y_0-y|\sin\alpha}\, d\alpha; \tag{5.61}$$

and, discarding the odd part of the integrand, this is

$$H_z^i = \frac{e^{-\frac{1}{4}i\pi}}{\sqrt{2\pi}} \int_{S(0)} \cos[k_0(y_0 - y)\sin\alpha]\, e^{-ik_0 d\cos\alpha}\, d\alpha. \tag{5.62}$$

The corresponding form for the reflected field is

$$H_z^r = \frac{e^{-\frac{1}{4}i\pi}}{2\sqrt{2\pi}} \int_{S(0)} \{[\varrho_H(\alpha) + \varrho_H(-\alpha)]\cos[k_0(y_0 + y)\sin\alpha] - i[\varrho_H(\alpha) - \varrho_H(-\alpha)]\sin[k_0(y_0 + y)\sin\alpha]\}\, e^{-ik_0 d\cos\alpha}\, d\alpha. \tag{5.63}$$

With the expression (5.8) inserted for ϱ_H, (5.62) and (5.63) in combination give the total field

$$H_z = \frac{e^{-\frac{1}{4}i\pi}}{\sqrt{2\pi}} \int_{S(0)} \Gamma(\alpha) \frac{2\sin^2\alpha}{\sin^2\alpha - \dfrac{1}{\mu^2}\left(1 - \dfrac{\cos^2\alpha}{\mu^2}\right)} e^{-ik_0 d\cos\alpha}\, d\alpha, \tag{5.64}$$

where

$$\Gamma(\alpha) = \cos(k_0 y_0 \sin\alpha)\cos(k_0 y \sin\alpha) - \frac{1}{\mu^2}\left(1 - \frac{\cos^2\alpha}{\mu^2}\right)\frac{\sin(k_0 y_0 \sin\alpha)\sin(k_0 y\sin\alpha)}{\sin^2\alpha} + \frac{i}{\mu}\sqrt{1 - \frac{\cos^2\alpha}{\mu^2}}\,\frac{\sin[k_0(y_0 + y)\sin\alpha]}{\sin\alpha}. \tag{5.65}$$

An asymptotic approximation to (5.64) for $k_0 d \gg 1$ is now obtained by putting $\alpha = 0$ in the factor $\Gamma(\alpha)$ in the integrand. Thus

$$H_z = \Gamma(0)\,(H_z)_{y_0 = y = 0}, \tag{5.66}$$

and the height-gain factor $\Gamma(0)$ is

$$(1 + ik_0 y_0 \sin\alpha_B)(1 + ik_0 y \sin\alpha_B), \tag{5.67}$$

where, as before, $\sin\alpha_B$ takes its approximate value $\sqrt{1 - 1/\mu^2}/\mu$.

The approximation (5.66) implies, of course, the "impedance" boundary condition

$$\frac{\partial H_z}{\partial y} = ik_0 H_z/\mu, \quad \text{for} \quad |\mu| \gg 1. \tag{5.68}$$

A rough criterion for the validity of the derivation of (5.66) is

$$k_0^2(y_0^2 + y^2) \ll k_0 d. \tag{5.69}$$

This states that the vertical excursions of source and point of observation must be so confined that both R and S differ from d by much less than a wavelength.

By taking d sufficiently large it is evidently possible to satisfy (5.69) and have at the same time $k_0 y|\sin\alpha_B| \gg 1$. Then, with $y_0 = 0$, application of the height-gain factor to (5.55) gives

$$\Delta = \frac{2y}{d\sin\alpha_B}, \tag{5.70}$$

which is in effect the same as (5.57) with $S = R$ and $\sin \psi = y/d \ll |\sin \alpha_B|$. On the other hand if also $k_0 y_0 |\sin \alpha_B| \gg 1$, the corresponding formula is

$$\Delta = 2iy_0 y/d, \tag{5.71}$$

which is implicit in (5.57) when the reflection coefficient is taken as -1.

One further observation is helpful in giving a vivid picture of the nature of the field. It can hardly be overlooked that the analysis of § 5.1.3, and the Fresnel integral form of the solution, are strongly reminiscent of the theory of diffraction by a black half-plane (§ 4.1.2). The present problem can, in fact, be expressed in terms of diffraction at an edge in the following way. In (4.16) replace α_0 by $\frac{1}{2}\pi - \alpha_B$, and θ by $\psi + 3\pi/2$, and compare the result with (5.45). Ignoring factors which are essentially unity when ψ and $|\alpha_B|$ are small, it appears that the two are identical apart from a constant factor. The field specified by (5.49) can therefore be described as the superposition on the field for a perfectly conducting half-space of the (vacuum) field due to the diffraction of the plane wave

$$H_z = -2\sqrt{2\pi}\, e^{\frac{1}{4}i\pi} \sin \alpha_B\, e^{-ik_0 S \cos(\psi + \alpha_B)} \tag{5.72}$$

by a black screen extending upwards from the image point of the line-source. The situation is illustrated in Fig. 5.5.

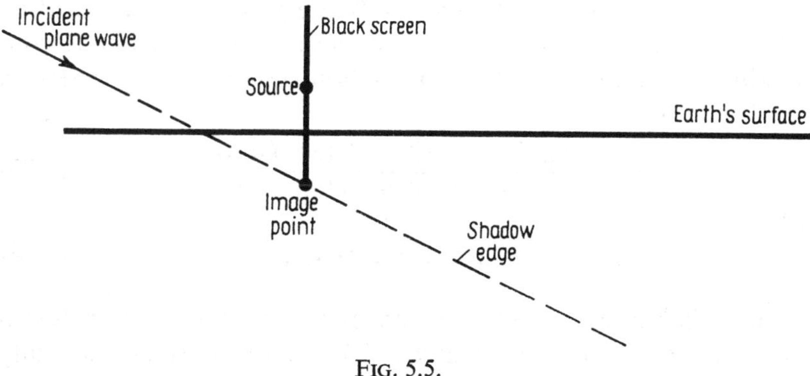

FIG. 5.5.

The plane wave (5.72) is incident at the Brewster angle, and the points of observation on or above the earth's surface are in the "shadow" of the screen, so that the plane wave is not seen directly.

The picture is the more readily envisaged when α_B is real, but the analysis has shown that no significant change takes place when α_B is complex, provided, as here, $\arg \alpha_B < \frac{1}{4}\pi$. A new feature in the behaviour of the solution would arise if it were possible to have $\arg \alpha_B > \frac{1}{4}\pi$; and the way in which this can, in effect, be achieved is the next subject for discussion.

5.2. SURFACE WAVES

5.2.1. Reactive Surfaces

Consider the H-polarized plane wave specified by

$$H_z^i = e^{ik_0 r \cos(\theta - \alpha)}$$

incident from the vacuum half-space $y > 0$ on the plane $y = 0$, below which is a medium whose properties are independent of x and z. Then in general the reflection coefficient will have poles at certain, possibly complex, values of α; suppose that they are determined by $\sin \alpha = -\sin \alpha_B$. Now it has been seen that if the properties of the medium are also independent of y, then, with $0 < \operatorname{Re} \alpha_B < \frac{1}{2}\pi$, the argument of α_B can only lie between 0 and $\frac{1}{4}\pi$. On the other hand, for stratified media this is by no means necessarily the case, as the following example shows.

Suppose the region between $y = 0$ and $y = -a$ is occupied by a homogeneous medium, and the plane $y = -a$ is perfectly conducting. Then an elementary calculation shows that the reflection coefficient is

$$\varrho_H = \frac{Z_0 \sin \alpha - iZ \sin \beta \tan (ka \sin \beta)}{Z_0 \sin \alpha + iZ \sin \beta \tan (ka \sin \beta)}, \tag{5.73}$$

where

$$k_0 \cos \alpha = k \cos \beta, \tag{5.74}$$

Z and k being respectively the impedance and propagation coefficient of the medium. Evidently (5.73) reduces correctly to unity when $a = 0$, and to (5.6) when $a \to \infty$ and k is complex, since in the latter case $\tan(ka \sin \beta) \to -i$.

To simplify the analysis, consider the quite practical case in which the coating on the perfectly conducting sheet is so thin that $|k|a \ll 1$; and let its specific permeability be unity, so that the relations (5.1), (5.2) and (5.3) hold. Then (5.73) is approximately

$$\varrho_H = \frac{\sin\alpha - ik_0 a \sin^2\beta}{\sin\alpha + ik_0 a \sin^2\beta}, \tag{5.75}$$

and (5.74) is

$$\cos\alpha = \mu \cos\beta. \tag{5.76}$$

The zeros and poles of ϱ_H are therefore given respectively by the relations $\sin\alpha = \sin\alpha_B$, $\sin\alpha = -\sin\alpha_B$, where

$$\sin^2\alpha_B + \frac{i\mu^2}{k_0 a}\sin\alpha_B + \mu^2 - 1 = 0. \tag{5.77}$$

Of the two solutions of (5.77), one tends to infinity as $a \to 0$, and is not important, and the other is approximately

$$\sin\alpha_B = ik_0 a(1 - 1/\mu^2). \tag{5.78}$$

Since

$$\mu^2 = (\varepsilon' - i\sigma/\omega)/\varepsilon_0,$$

where ε' and σ are the permittivity and conductivity of the medium, (5.78) gives

$$\tan[\arg(\sin\alpha_B)] = \frac{\varepsilon'(\varepsilon' - \varepsilon_0) + \sigma^2/\omega^2}{\varepsilon_0\sigma/\omega}. \tag{5.79}$$

Thus the argument of $\sin\alpha_B$ lies between some positive minimum value and $\frac{1}{2}\pi$, being equal to $\frac{1}{2}\pi$ when $\sigma = 0$.

A further simplifying approximation is afforded by the case $|\mu| \gg 1$. Then for most purposes, as indicated for example by the analysis of § 5.1.3, it is legitimate to replace $\sin\beta$ by unity in (5.75); thus

$$\varrho_H = \frac{\sin\alpha - ik_0 a}{\sin\alpha + ik_0 a}, \tag{5.80}$$

which, of course, demands the neglect of $1/\mu^2$ compared with unity in (5.78). It is an implication of (5.80) that it is permissible to treat the plane $y = 0$ as an impedance surface, in the sense that the field in $y > 0$ can be obtained simply by applying the boundary condition

$$E_x = Z_S H_z \quad \text{at} \quad y = 0, \tag{5.81}$$

without need to refer to the field in $y < 0$ (cf. (5.32) and (5.68)). For (5.81) leads to

$$\varrho_H = \frac{Z_0 \sin\alpha - Z_S}{Z_0 \sin\alpha + Z_S}, \tag{5.82}$$

and comparison with (5.80) gives the identification

$$Z_S = ik_0 a Z_0. \tag{5.83}$$

Here, then, the surface impedance Z_S is moreover purely reactive.

The dielectric coated perfectly conducting sheet has been given as a particularly simple illustration, but there are also other methods of producing impedance surfaces which can be regarded as purely reactive, with reactance independent of the incident field. It is therefore not unrealistic to investigate the implications of a reflection coefficient of the form

$$\varrho_H = \frac{\sin\alpha - \sin\alpha_B}{\sin\alpha + \sin\alpha_B}, \tag{5.84}$$

where $\alpha_B = i\gamma$ is on the positive imaginary axis. The problem is of special interest, because the surface can evidently support the pure surface wave

$$H_z = e^{ik_0 r\cos(\theta - \alpha_B)} = e^{-k_0 y\sinh\gamma}\, e^{ik_0 x\cosh\gamma}. \tag{5.85}$$

The way in which such a wave appears as an important constituent of the total field generated by a source is now investigated by the method of § 5.1.3.

5.2.2. Generation of a Surface Wave

In this section the problem treated in § 5.1.3 is re-examined for the case when ϱ_H is specified not by (5.42) but rather by (5.84), with $\alpha_B = i\gamma$, where γ is real and positive; that is

$$\varrho_H(\alpha) = 1 - \frac{2i\sinh\gamma}{\sin\alpha + i\sinh\gamma}. \tag{5.86}$$

The incident field (5.40) thus gives rise to the total field (5.43), where now

$$H_{1z} = -\sqrt{\frac{2}{\pi}}\, e^{\frac{1}{4}i\pi}\sinh\gamma \int\limits_C \frac{1}{\sin\alpha + i\sinh\gamma}\, e^{ik_0 S\cos(\psi+\alpha)}\, d\alpha. \tag{5.87}$$

As before, the path C in (5.87) is distorted into the path of steepest descents, $S(\pi - \psi)$. The only singularities of the integrand are simple poles where $\sin\alpha = -i\sinh\gamma$; those at $-i\gamma$ and $\pi + i\gamma$ lie in principle just outside the strip $0 \leqq \operatorname{Re}\alpha \leqq \pi$, and it is clear that the latter is captured when ψ is sufficiently close to zero. More precisely, the pole is captured when $\psi < \psi_S$, where ψ_S is determined by the condition that $S(\pi - \psi_S)$ passes through $\pi + i\gamma$, and for present purposes need not be particularized further. Thus

$$H_{1z} = -\sqrt{\frac{2}{\pi}}\, e^{\frac{1}{4}i\pi} \sinh\gamma \int\limits_{S(\pi-\psi)} \frac{1}{\sin\alpha + i\sinh\gamma}\, e^{ik_0 S\cos(\psi+\alpha)}\, d\alpha$$

$$+ 2\sqrt{2\pi}\, e^{-\frac{1}{4}i\pi} \tanh\gamma\, e^{-k_0(y+y_0)\sinh\gamma}\, e^{-ik_0 x\cosh\gamma}\, U(\psi - \psi_S), \quad (5.88)$$

where U is the step-function that is zero or unity according to whether its argument is positive or negative. In this expression the integral term is simply (5.87) with C replaced by $S(\pi - \psi)$, and the surface wave term is the contribution from the residue of the pole at $\alpha = \pi + i\gamma$; y_0 and y are the heights of source and point of observation, respectively, above the reactive surface, and x is distance along the surface between source and point of observation.

If desired, the integral term in (5.88) can be approximated by a Fresnel integral in much the same manner as was the corresponding expression in § 5.1.3; it need only be noted that the sign of the argument of the Fresnel integral depends on whether ψ is greater or less than ψ_S, as indicated in (3.65), the resulting discontinuity at $\psi = \psi_S$ of course counterbalancing that associated with the appearance of the surface wave term. Here, however, the main interest is in the contrasting character of the two parts of the field, and this is shown most clearly by letting S become very large. Then the surface wave term in (5.88) persists to very large values of y, and the integral term is of order $1/\sqrt{S}$. In this sense the total field can be regarded as the superposition of a "space" wave and a surface wave. For a fixed value of y and indefinitely increasing x, the former, as has been seen, is of order $1/x^{3/2}$, whereas the latter suffers no reduction in amplitude. At points close to the surface the contribution of the surface wave therefore dominates that of the "space" wave.

In practice the surface wave must exhibit some attenuation in the x-direction resulting from the fact that the surface is not a pure reactance. Nevertheless it remains true in application that power can be usefully transmitted by surface waves, and it is of interest to examine the efficiency of the technique in the ideal case. To this end some further development of the theory is given in the next section.

5.2.3. Launching Efficiency

On the assumption that a purely reactive, infinite, plane surface is available, the question arises as to what fraction of the power emitted by a source goes into the surface wave. The fraction is known as the *launching efficiency*, and depends on the nature of the source. Perhaps the commonest situation envisaged, in the context of a two-dimensional plane geometry, is that in which the source is represented by a prescribed distribution of vertical electric field, E_y, over a vertical plane, say $x = 0$. The special case for which the distribution is a delta function, the voltage slot (see (2.99) and (2.101)), is indeed just that given in § 5.2.2. The general case, the "voltage aperture", is equivalent to the superposition of voltage slots; but rather than integrate the previous solution over y_0 it is more convenient once again to begin by representing the undisturbed field of the aperture distribution as an angular spectrum of plane waves.

Only one point prevents the nature of the generalization from being entirely trivial; namely that the aperture distribution is at right angles to the reactive plane. Whereas the most natural representation of the undisturbed source field is in terms of plane waves fanning into the entire half-space $x > 0$, it is necessary, in order immediately to write down correctly the corresponding reflected field, to use a representation for the half-space $y < 0$; this ensures that each plane wave of the spectrum is indeed incident on the surface, and leaves no room for doubt about whether or not conditions at infinity have been properly taken into account.

Once the requirement is understood it is easy to put into effect. Suppose that the voltage aperture in the plane $x = 0$ extends from $y = 0$ up to some finite height $y = h$; then $E_y^i(0, y)$ (where affix i, as before, denotes the undisturbed source field) is prescribed, and in particular is zero for both $y < 0$ and for $y > h$. If it is imagined that this is achieved by the aperture carrying the appropriate surface *magnetic* current distribution, as discussed in the two paragraphs following (2.84), then the source currents are all above $y = 0$, and the undisturbed source field can therefore be represented analytically throughout the half-space $y < 0$.

Before setting out the mathematical expressions it is worth making one further observation. For the magnetic current aperture, $E_y^i(x, y)$ is antisymmetric in x; the corresponding component of the reflected field, $E_y^r(x, y)$, is likewise antisymmetric in x, and therefore zero on the plane $x = 0$. Thus the total field is such that $E_y(0, y) = E_y^i(0, y)$, and the solution to be given is in effect the H-polarized field in the quarter space $x > 0$, $y > 0$ that satisfies the impedance boundary condition on $y = 0$ and has prescribed behaviour of E_y on $x = 0$.

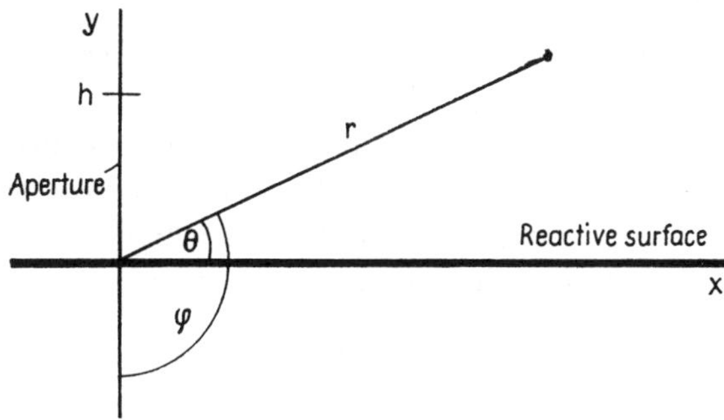

FIG. 5.6.

It is perhaps simplest to introduce explicitly the two angles θ, φ, as shown in Fig. 5.6, with $\varphi = \frac{1}{2}\pi + \theta$. Then our conventional representation of the undisturbed field of the aperture distribution in $x = 0$ can be written

$$H_z^i = \int_C P(\cos\beta)\, e^{-ik_0 r\cos(\varphi-\beta)}\, d\beta, \tag{5.89}$$

valid for $0 < \varphi < \pi$. The corresponding expression for E_y^i merely has the extra factor $Z_0 \sin\beta$ in the integrand; putting $x = 0$ and $\cos\beta = \lambda$, inverting the Fourier transform, and recalling that $E_y^i(0, y)$ is zero unless $0 < y < h$, gives

$$P(\lambda) = \frac{k_0 Y_0}{2\pi} \int_0^h E_y^i(0, y)\, e^{-ik_0\lambda y}\, dy. \tag{5.90}$$

These formulae are nothing other than (2.77) and (2.84) with appropriate changes of notation. $P(\lambda)$ is determined by the prescription of $E_y^i(0, y)$, and is free of singularities by virtue of the finite range of integration.

If now the substitutions $\varphi = \frac{1}{2}\pi + \theta$, $\beta = \alpha - \frac{1}{2}\pi$ are made in (5.89), it reads

$$H_z^i = \int_{C+\frac{1}{2}\pi} P(\sin\alpha)\, e^{ik_0 r\cos(\theta-\alpha)}\, d\alpha, \tag{5.91}$$

valid for $-\frac{1}{2}\pi < \theta < \frac{1}{2}\pi$, where the path $C + \frac{1}{2}\pi$ is the path C translated a distance $\frac{1}{2}\pi$ parallel to the real axis. The analytic continuation of this field into the entire half-space $-\pi < \theta < 0$ is now obvious, being simply (5.91) with $C + \frac{1}{2}\pi$ replaced by C, and the analysis can proceed as a trivial generalization of that in § 5.2.2.

The field reflected into $0 < \theta < \frac{1}{2}\pi$ is

$$H_z^r = \int_C \varrho_H(\alpha)\, P(\sin\alpha)\, e^{ik_0 r\cos(\theta+\alpha)}\, d\alpha, \tag{5.92}$$

with ϱ_H given by (5.86). The path C can be distorted into the steepest descents path $S(\pi-\theta)$, with allowance necessary only for the pole of ϱ_H at $\alpha = \pi + i\gamma$ when $\theta < \theta_S$, where θ_S is the angle for which $S(\pi-\theta_S)$ passes through the pole. Thus

$$H_z^r = \int_{S(\pi-\theta)} \frac{\sin\alpha - i\sinh\gamma}{\sin\alpha + i\sinh\gamma}\, P(\sin\alpha)\, e^{ik_0 r\cos(\theta+\alpha)}\, d\alpha$$
$$+\, 4\pi\tanh\gamma\, P(-i\sinh\gamma)\, e^{-k_0 y\sinh\gamma}\, e^{-ik_0 x\cosh\gamma}\, U(\theta-\theta_S), \tag{5.93}$$

where U is the step-function that is zero or unity according to whether its argument is positive or negative, and, from (5.90),

$$P(-i\sinh\gamma) = \frac{k_0 Y_0}{2\pi}\int_0^h E_y^i(0, y)\, e^{-k_0 y\sinh\gamma}\, dy. \tag{5.94}$$

These formulae put in evidence the separation of the total field into "space" wave and surface wave components in the manner already described in § 5.2.2. The discussion is now completed by considering the power associated with each component. The concept can be made quite definite by imagining r to tend to infinity. The surface wave then extends upwards to an unlimited height, and its time-averaged power flux in the x-direction, per unit length in the z-direction, is (cf. (2.24))

$$16\pi^2\tanh^2\gamma\, |P(-i\sinh\gamma)|^2\, \tfrac{1}{2}Z_0\cosh\gamma \int_0^\infty e^{-2k_0 y\sinh\gamma}\, dy$$
$$= \frac{4\pi^2 Z_0}{k_0}\tanh\gamma\, |P(-i\sinh\gamma)|^2. \tag{5.95}$$

The magnetic field of the space wave, which is the superposition, in $0 < \theta < \frac{1}{2}\pi$, of H_z^i and the integral term of (5.93), can be written

$$\int_{S(\theta)} \left[P(-\sin\alpha) + \frac{\sin\alpha - i\sinh\gamma}{\sin\alpha + i\sinh\gamma}\, P(\sin\alpha)\right] e^{-ik_0 r\cos(\theta-\alpha)}\, d\alpha. \tag{5.96}$$

This is precisely the form discussed in § 3.2. The corresponding radiation field is

$$\sqrt{2\pi}\, e^{\frac{1}{4}i\pi}\left[P(-\sin\theta) + \frac{\sin\theta - i\sinh\gamma}{\sin\theta + i\sinh\gamma}P(\sin\theta)\right]\frac{e^{-ik_0 r}}{\sqrt{k_0 r}}, \tag{5.97}$$

and the power radiated into $0 < \theta < \frac{1}{2}\pi$ is

$$\frac{\pi Z_0}{k_0}\int_0^{\frac{1}{2}\pi}\left|P(-\sin\theta) + \frac{\sin\theta - i\sinh\gamma}{\sin\theta + i\sinh\gamma}P(\sin\theta)\right|^2 d\theta. \tag{5.98}$$

It may be questioned whether the total power flux across the aperture plane is the sum of (5.95) and (5.98). One way of proving that this is so is to calculate the former directly. The change of variable $\alpha = 3\pi/2 - \beta$ in (5.92), followed by translation of the β path of integration, which is initially $C + \frac{1}{2}\pi$, back to C, thus crossing the pole, gives the total field as

$$H_z = \int_C \left[P(\cos\beta) + \frac{\cos\beta + i\sinh\gamma}{\cos\beta - i\sinh\gamma}P(-\cos\beta)\right]e^{-ik_0 r\cos(\varphi-\beta)}\,d\beta$$
$$+\,4\pi\tanh\gamma\, P(-i\sinh\gamma)\, e^{-k_0 y\sinh\gamma}\, e^{-ik_0 x\cosh\gamma}. \tag{5.99}$$

With $x = 0$, the conjugate complex of this expression is now multiplied by $\frac{1}{2}E_y^i(0, y)$, the product integrated over y from 0 to h, and the real part taken. It follows at once from (5.94) that the contribution associated with the last term of (5.99) is precisely (5.95). The contribution from the integral in (5.99), after the y integration has been identified with (5.90), appears as

$$\mathrm{Re}\,\frac{\pi Z_0}{k_0}\int_C P(\cos\beta)\left[P^*(\cos\beta) + \frac{\cos\beta - i\sinh\gamma}{\cos\beta + i\sinh\gamma}P^*(-\cos\beta)\right]d\beta.$$

If from this expression an equivalent one is derived by replacing β by $\pi - \beta$, and half the sum of the two taken, the result is

$$\frac{\pi Z_0}{2k_0}\int_0^{\pi}\left|P(\cos\beta) + \frac{\cos\beta + i\sinh\gamma}{\cos\beta - i\sinh\gamma}P(-\cos\beta)\right|^2 d\beta, \tag{5.100}$$

where the parts of C off the real axis are discarded when the real part is taken since on them $d\beta$ is purely imaginary. Evidently (5.100) is equivalent to (5.98).

CHAPTER VI

PROPAGATION OVER A TWO-PART PLANE SURFACE

6.1. PERFECTLY CONDUCTING HALF-PLANE ON SURFACE OF SEMI-INFINITE HOMOGENEOUS MEDIUM

6.1.1. Genesis and Nature of the Problem

In Chapter 5 the problem of propagation over a uniform surface was treated. A generalization, of both physical and mathematical interest, is that in which the surface consists of two semi-infinite sections with differing electrical properties. This model, for example, reproduces some of the effects associated with a path for radio propagation which is, say, partly over land and partly over sea; at sufficiently long wavelengths the dominant effects can be those from the high conductivity of the sea compared with land, the detailed nature of the coastline being relatively unimportant. Again, in the transmission of surface waves over a reactive surface, the model can be used to investigate the amount of reflection and radiation associated with an abrupt change in the reactance.

The problem is evidently likely to be rather more complicated than those already treated, and it is helpful first to consider a form which brings out particularly clearly the relation to the work presented both in Chapter 4 and Chapter 5. Imagine, then, that as in § 5.1 the half-space $y > 0$ is vacuum and the half-space $y < 0$ is occupied by some homogeneous medium; and further, that in the interface $y = 0$ there lies a perfectly conducting half-plane $x > 0$ (see Fig. 6.1). Although in the ensuing investigation of this boundary value problem emphasis is on the propagation aspects just outlined, which represent its main field of application, nevertheless it may also fairly be regarded as a direct generalization of the half-plane diffraction problems of §§ 4.2.2 and 4.2.5, to which it reverts if the medium in $y < 0$ is itself vacuum. Indeed, the isolated, perfectly conducting half-plane typifies most simply the "two-part" problem; as has been seen, it can be described in terms of a function $\varphi(x, y)$, say, that satisfies the two-dimensional time-harmonic wave equation, and has boundary conditions on $y = 0$, $\varphi = \varphi_0(x)$ (given) for $x > 0$, and $\partial\varphi/\partial y = 0$ for $x < 0$. The solutions now to be developed, first for an incident plane wave, and then for a line-source, follow precisely the methods of §§ 4.2.2, 4.2.5; the complication arises from the presence in the plane wave spectrum function of factors accounting for the reflection properties of the medium that occupies $y < 0$.

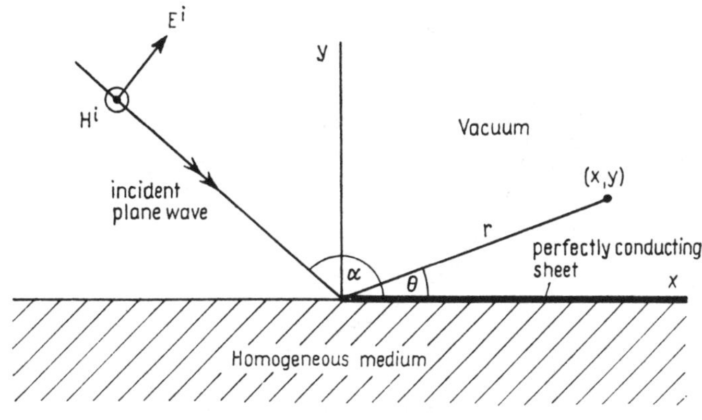

FIG. 6.1.

Before proceeding with the mathematics it is desirable briefly to make explicit a prominent feature of the problem, which is used ultimately as the main illustration of the character of the solution. Suppose, for definiteness, that a vertically polarized transmitter is situated at sea-level on flat, dry ground of poor conductivity, and that its distance from the coast is many "numerical distances" (§ 5.1.4). Imagine the vertical component of electric field measured at sea-level by a receiver which proceeds in a straight line away from the transmitter, first over land and subsequently over the sea. What might be expected in a plot of field-strength against distance from the transmitter? It is to be expected, for points not very close to the coast, that the field-strength over land will be unaffected by the presence of the sea, so that formula (5.55) is for the most part applicable. It might further be thought that, with the neglect of some secondary local disturbance in the vicinity of the coastline, the field-strength would continue to decrease with distance as the receiver crossed over to the sea, albeit possibly at a reduced rate because of the greatly improved conductivity. However, the latter supposition is quite false. This can be surmised by envisaging the interchange of transmitter and receiver, so that the transmitter is out to sea and the receiver on land. Then the above naïve argument suggests that the field-strength would be much enhanced, whereas it is known from the general reciprocity theorem that it is unaffected by the interchange. The point is made graphically in Fig. 6.2;

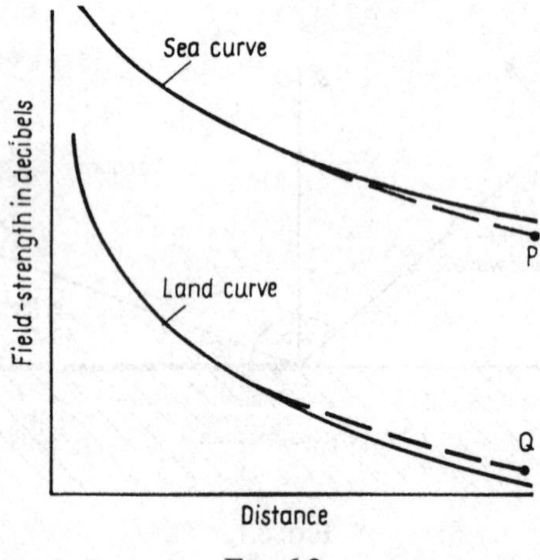

FIG. 6.2.

the full-line curves are the plots of field-strength, against distance from a given transmitter, that would be obtained were the propagation path all sea or all land, respectively, and the dashed curves represent the assumption that there is a continued decrease beyond the coastline at the rate appropriate to the new medium; the values indicated by P and Q could in practice be quite 20 decibels apart.

In fact, with the transmitter on land, the plot of the field-strength as the receiver moves away across the coastline shows an initial marked rise; it reaches a maximum appreciably above its value at the coastline, and only subsequently falls off at a rate appropriate to transmission over sea, ultimately settling down to a curve somewhere between the "all sea" and "all land" curves. It is especially this field-strength "recovery effect" that will be made explicit in the examination of the solution in § 6.1.5.

6.1.2. Solution for Incident Plane Wave: *H*-polarization

The two-dimensional problem with an H-polarized field is treated. To begin with, a solution is obtained for the case of the plane wave

$$H_z^i = e^{ik_0 r\cos(\theta-\alpha)}, \quad 0 \leqq \alpha \leqq \pi, \tag{6.1}$$

incident from $y > 0$, the coordinate system being shown in Fig. 6.1. The solution for a line-source is derived in § 6.1.3 by an integration over α.

If the perfectly conducting sheet were absent, (6.1) would give rise to the reflected wave

$$H_z^r = \varrho_H\, e^{ik_0 r\cos(\theta+\alpha)} \tag{6.2}$$

in $y > 0$, and the transmitted wave

$$H_z^t = \tau_H\, e^{ikr\cos(\theta-\alpha')} \tag{6.3}$$

in $y < 0$, where the reflection coefficient is

$$\varrho_H = \frac{\sin\alpha - \sin\alpha'/\mu}{\sin\alpha + \sin\alpha'/\mu}, \tag{6.4}$$

the transmission coefficient is

$$\tau_H = \frac{2\sin\alpha}{\sin\alpha + \sin\alpha'/\mu}, \tag{6.5}$$

and

$$\cos\alpha = \mu\cos\alpha'. \tag{6.6}$$

The notation is the same as that used in § 5.1.1, with $\mu = k/k_0$ denoting the refractive index of the medium, except that β (being required elsewhere) has been replaced by α'.

With the perfectly conducting sheet present there will be, in addition to the above fields, a scattered field generated by the surface currents in the sheet. This scattered field is, as usual, expressed as a spectrum of plane waves. For $y > 0$, the now familiar form

$$H_z^s = \int_C P(\cos\beta)\, e^{-ik_0 r\cos(\theta-\beta)}\, d\beta, \tag{6.7}$$

with correspondingly

$$E_x^s = -Z_0 \int_C \sin\beta\, P(\cos\beta)\, e^{-ik_0 r\cos(\theta-\beta)}\, d\beta, \tag{6.8}$$

is appropriate. For $y < 0$, account can be straightway taken of the fact that E^s_x is continuous across $x = 0$ by writing

$$H^s_z = -\mu \int_C \frac{\sin\beta}{\sin\beta'} P(\cos\beta)\, e^{-ik_0 x\cos\beta}\, e^{iky\sin\beta'}\, d\beta, \qquad (6.9)$$

$$E^s_x = -Z_0 \int_C \sin\beta\, P(\cos\beta)\, e^{-ik_0 x\cos\beta}\, e^{iky\sin\beta'}\, d\beta, \qquad (6.10)$$

where

$$\cos\beta = \mu\cos\beta'. \qquad (6.11)$$

Thus although the scattered field has no simple symmetry about $y = 0$ comparable to that in the isolated half-plane problem, it is readily expressed in terms of a single spectrum function $P(\cos\beta)$, which function is to be determined from the remaining boundary conditions at $y = 0$. For the total field **E**, **H** specified by

$$H_z = \begin{cases} H^i_z + H^r_z + H^s_z, & \text{for } y \geqq 0, \\ H^t_z + H^s_z, & \text{for } y \leqq 0, \end{cases} \qquad (6.12)$$

these are, first, continuity of H_z at $y = 0$, $x < 0$, since there is, of course, no surface current apart from that on the perfectly conducting sheet; and secondly, vanishing of E_x at $y = 0$, $x > 0$. But at $y = 0$, $H^i_z + H^r_z = H^t_z$ and $E^i_x + E^r_x = E^t_x$; the boundary conditions are therefore equivalent to

$$H^s_z \text{ is continuous at } y = 0, \quad x < 0, \qquad (6.13)$$

$$E^s_x = -E^t_x \text{ at } y = 0, \quad x > 0. \qquad (6.14)$$

From the representations (6.7) and (6.9) with $y = 0$, writing $\lambda_0 = \cos\alpha$ and making the change of integration variable

$$\lambda = \cos\beta, \qquad (6.15)$$

(6.13) becomes

$$\int_{-\infty}^{\infty} \frac{\sqrt{1-\lambda^2} + \sqrt{1-\lambda^2/\mu^2}/\mu}{\sqrt{1-\lambda^2}\,\sqrt{1-\lambda^2/\mu^2}} P(\lambda)\, e^{-ik_0 x\lambda}\, d\lambda = 0, \quad \text{for} \quad x < 0; \qquad (6.16)$$

and similarly the representations (6.8) and (6.10) give (6.14) as

$$\int_{-\infty}^{\infty} P(\lambda)\, e^{-ik_0 x\lambda}\, d\lambda = \frac{2}{\mu} \frac{\sqrt{1-\lambda_0^2}\,\sqrt{1-\lambda_0^2/\mu^2}}{\sqrt{1-\lambda_0^2} + \sqrt{1-\lambda_0^2/\mu^2}/\mu} e^{ik_0 x\lambda_0}, \quad \text{for} \quad x > 0. \qquad (6.17)$$

Evidently (6.16), (6.17) are dual integral equations for $P(\lambda)$. In the case $\mu = 1$, they revert to the equations for the corresponding isolated half-plane problem, being then quite similar to the E-polarization equations (4.48), (4.49); and the method of solution is, in principle, no different from that described in § 4.2.2. The λ path of integration avoids the branch-points at ± 1 in the way shown in Fig. 2.2, and the objective is to close the path with infinite semicircles, above the real axis in (6.16), since $x < 0$, and below the real axis in (6.17), since $x > 0$. As before, the concept of U and L functions is introduced, and solutions of (6.16) and (6.17), respectively, are written in the form

$$\frac{\sqrt{1-\lambda^2} + \sqrt{1-\lambda^2/\mu^2}/\mu}{\sqrt{1-\lambda^2}\,\sqrt{1-\lambda^2/\mu^2}}\, P(\lambda) = U(\lambda), \tag{6.18}$$

$$P(\lambda) = -\frac{1}{2\pi i}\,\frac{2}{\mu}\,\frac{\sqrt{1-\lambda_0^2}\,\sqrt{1-\lambda_0^2/\mu^2}}{\sqrt{1-\lambda_0^2} + \sqrt{1-\lambda_0^2/\mu^2}/\mu}\,\frac{L(\lambda)}{L(-\lambda_0)\,(\lambda+\lambda_0)}, \tag{6.19}$$

where the path of integration is indented above the pole at $\lambda = -\lambda_0$.

From the elimination of $P(\lambda)$ between (6.18) and (6.19) it is clear that the solution of the equations requires the expression of

$$\sqrt{1-\lambda^2} + \sqrt{1-\lambda^2/\mu^2}/\mu \tag{6.20}$$

as the product of a U function and an L function. Now it is known that any function can be "split" in this way, but the explicit factors are liable to be complicated, especially if branch-points are present; the example $\sqrt{1-\lambda^2}$ arising in the isolated half-plane problem is exceptional in its simplicity. Here we shall be content merely to accept that the factorization is possible in principle, because this proves sufficient to give the main results. Moreover, as shown in § 5, in the common case $|\mu| \gg 1$ it is permissible to replace the second term in (6.20) by the constant $\sqrt{1-1/\mu^2}/\mu$. The split of the resulting simpler function, though by no means trivial, can be achieved by comparatively elementary methods, and this is done in § 6.2.2 in connection with an analogous surface wave problem.

As a formal expression of the factorization, write

$$\frac{2}{\mu}\,\frac{\sqrt{1-\lambda^2/\mu^2}}{\sqrt{1-\lambda^2} + \sqrt{1-\lambda^2/\mu^2}/\mu} = \frac{1}{U_1(\lambda)\,L_1(\lambda)}. \tag{6.21}$$

Note that $L_1(-\lambda)$ must be a constant multiple of $U_1(\lambda)$, which constant it is obviously permissible and convenient to take to be unity; and that $U_1(\lambda) = L_1(\lambda) = 1$ if $\mu = 1$. Then from (6.18) and (6.19),

$$P(\lambda) = -\frac{1}{2\pi i}\,\frac{\sqrt{1+\lambda_0}\,\sqrt{1+\lambda}}{L_1(\lambda_0)\,L_1(\lambda)}\,\frac{1}{\lambda+\lambda_0}; \tag{6.22}$$

or, equivalently,

$$P(\cos\beta) = \frac{i}{\pi}\,\frac{1}{L_1(\cos\alpha)\,L_1(\cos\beta)}\,\frac{\cos(\tfrac{1}{2}\alpha)\cos(\tfrac{1}{2}\beta)}{\cos\alpha+\cos\beta}. \tag{6.23}$$

Reciprocity is manifested by the symmetry of (6.23) in α and β (cf. the remarks at the end of § 4.2.5).

From now on only the field in $y \geqq 0$ is considered. The substitution of (6.23) into (6.7) gives the field generated by the currents in the perfectly conducting half-plane, so the total field is given by

$$H_z = e^{ik_0 r\cos(\theta-\alpha)} + \varrho_H\, e^{ik_0 r\cos(\theta+\alpha)} + \frac{i}{\pi}\,\frac{\cos(\tfrac{1}{2}\alpha)}{L_1(\cos\alpha)} \times \int_C \frac{\cos(\tfrac{1}{2}\beta)}{L_1(\cos\beta)(\cos\beta+\cos\alpha)}\, e^{-ik_0 r\cos(\theta-\beta)}\, d\beta. \tag{6.24}$$

The aim is not to examine this solution in detail, but rather to use it to find the solution for a line-source by integrating over α precisely in the manner of § 4.2.5. To this end the expression of (6.24) in the form of the superposition of the "geometrical optics" field and the "diffraction" field is required, which means analytically the distortion of the path C into the path of steepest descents $S(\theta)$, with due allowance for the singularities of the integrand.

Now it is recalled from § 5.1 that (6.21), regarded as a function of β, has the singularities shown in Fig. 5.4 (now reading β for α). Thus $L_1(\cos\beta)$ has branch-points at $\cos\beta = -\mu$, and a pole at $\beta = \pi + \alpha_B$, where α_B is the (complex) Brewster angle defined in § 5.1.1; the branch-points at $\cos\beta = \mu$ and the pole at $\beta = -\alpha_B$ belong to $U_1(\cos\beta)$. The implication is that if θ is near π the singularities of $L_1(\cos\beta)$ have a relevance to the evaluation of (6.24) quite analogous to that of the singularities of the integrand in the problem of a line-source over a homogeneous earth discussed in § 5.1.3; whereas if θ is near zero the singularities are of little significance. This is to be expected, since in the former case the disturbance associated with the edge $x = y = 0$ of the half-plane has travelled over the medium, whereas in the latter case it has travelled over a perfect conductor. In any event, the only pole of the integrand of (6.24) that can be captured by the path distortion is that at $\beta = \pi - \alpha$, and the branch-cut contribution that might strictly have to be included will be neglected on the basis of the arguments given in § 5.1.3. It is then readily seen that, for $y \geqq 0$,

$$H_z = H_z^g + H_z^d, \tag{6.25}$$

where

$$H_z^g = \begin{cases} e^{ik_0 r \cos(\theta-\alpha)} + e^{ik_0 r \cos(\theta+\alpha)}, & \text{for } 0 \leqq \theta \leqq \pi - \alpha, \\ e^{ik_0 r \cos(\theta-\alpha)} + \varrho_H\, e^{ik_0 r \cos(\theta+\alpha)}, & \text{for } \pi - \alpha \leqq \theta \leqq \pi, \end{cases} \tag{6.26}$$

and

$$H_z^d = \frac{i}{\pi}\frac{\cos(\frac{1}{2}\alpha)}{L_1(\cos\alpha)} \int\limits_{S(\theta)} \frac{\cos(\frac{1}{2}\beta)}{L_1(\cos\beta)(\cos\beta + \cos\alpha)}\, e^{-ik_0 r\cos(\theta-\beta)}\, d\beta\,. \tag{6.27}$$

6.1.3. Solution for Line-source: *H*-polarization

Consider a line-source situated at (r_0, θ_0) and specified by (4.90). As in § 4.2.5 the solution is obtained by multipling (6.25) by (4.93) and integrating over α along the path $S(\theta_0)$. The results analogous to (4.96), (4.97) and (4.98) can evidently be written

$$H_z = H_z^g + H_z^d, \tag{6.28}$$

where

$$H_z^g = \begin{cases} \sqrt{\tfrac{1}{2}\pi}\, e^{-\frac{1}{4}i\pi}[H_0^{(2)}(k_0R) + H_0^{(2)}(k_0S)], \quad \text{for } 0 \leqq \theta \leqq \pi - \theta_0, \\ \sqrt{\tfrac{1}{2}\pi}\, e^{-\frac{1}{4}i\pi} H_0^{(2)}(k_0R) + \dfrac{e^{-\frac{1}{4}i\pi}}{\sqrt{2\pi}} \displaystyle\int\limits_{S(\frac{1}{2}\pi)} \varrho_H\, e^{ik_0 S\cos(\psi+\alpha)}\, d\alpha, \\ \qquad\qquad \text{for } \pi - \theta_0 \leqq \theta \leqq \pi, \end{cases} \tag{6.29}$$

and

$$H_z^d = \frac{e^{\frac{1}{4}i\pi}}{\pi\sqrt{2\pi}} \int\limits_{S(\theta_0)}\int\limits_{S(\theta)} \frac{\cos(\frac{1}{2}\alpha)\cos(\frac{1}{2}\beta)}{L_1(\cos\alpha)\, L_1(\cos\beta)\,(\cos\alpha + \cos\beta)}$$
$$\times\, e^{-ik_0[r_0\cos(\theta_0-\alpha) + r\cos(\theta-\beta)]}\, d\beta\, d\alpha\,. \tag{6.30}$$

The symbols R, S and ψ have the same meaning as in § 5.1, and the "geometrical optics" field (6.29) can be expressed in terms of (5.44) as

$$H_z^g = \begin{cases} \sqrt{\tfrac{1}{2}\pi}\, e^{-\frac{1}{4}i\pi}[H_0^{(2)}(k_0R) + H_0^{(2)}(k_0S)], & \text{for } 0 \leqq \theta \leqq \pi - \theta_0, \\ \sqrt{\tfrac{1}{2}\pi}\, e^{-\frac{1}{4}i\pi}[H_0^{(2)}(k_0R) + H_0^{(2)}(k_0S)] + H_{1z}, & \\ & \text{for } \pi - \theta_0 \leqq \theta \leqq \pi. \end{cases} \tag{6.31}$$

The "diffraction" field (6.30) must counterbalance the discontinuity in (6.31), and is obviously of importance in the vicinity of the line $\theta = \pi - \theta_0$. Since in practical situations $\theta + \theta_0$ may well be close to π, it is evident that the crux of the problem is the reduction of the double integral in (6.30) to a tractable form.

It is assumed that both k_0r_0 and k_0r are large, and, by virtue of the exponential factor, the integrand of (6.30) then contributes significantly to the value of the integral only when α and β are near θ_0 and θ respectively. A now familiar type of asymptotic approximation can therefore be obtained by leaving under the integral sign no more of the non-exponential part of the integrand than is necessary to retain the poles that may lie close to $\alpha = \theta_0$ or $\beta = \theta$. A suitable form for the approximation is

$$\begin{aligned} H_z^d &= \frac{e^{\frac{1}{4}i\pi}}{\pi\sqrt{2\pi}} \frac{1}{L_2(\cos\theta_0)\, L_2(\cos\theta)} \\ &\times \int\limits_{S(\theta_0)}\int\limits_{S(\theta)} \frac{\cos(\frac{1}{2}\alpha)\cos(\frac{1}{2}\beta)}{\cos\left(\dfrac{\alpha-\alpha_B}{2}\right)\cos\left(\dfrac{\beta-\alpha_B}{2}\right)(\cos\alpha + \cos\beta)} \\ &\times e^{-ik_0[r_0\cos(\theta_0-\alpha)+r\cos(\theta-\beta)]}\, d\beta\, d\alpha, \end{aligned} \tag{6.32}$$

where

$$L_2(\cos\alpha) = \sec\left(\frac{\alpha-\alpha_B}{2}\right) L_1(\cos\alpha); \tag{6.33}$$

and (6.32) can in turn conveniently be written (cf. (4.99))

$$H_z^d = \frac{e^{\frac{1}{4}i\pi}}{4\pi\sqrt{2\pi}} \frac{I_1 + I_2}{L_2(\cos\theta_0)\, L_2(\cos\theta)}, \tag{6.34}$$

where

$$\begin{aligned} I_1 &= \int\limits_{S(0)}\int\limits_{S(0)} \\ &\times \frac{e^{-ik_0(r_0\cos\alpha + r\cos\beta)}}{\cos\left(\dfrac{\alpha+\theta_0-\alpha_B}{2}\right)\cos\left(\dfrac{\beta+\theta-\alpha_B}{2}\right)\cos\left(\dfrac{\alpha-\beta+\theta_0-\theta}{2}\right)} \\ &\times d\alpha\, d\beta, \end{aligned} \tag{6.35}$$

$$I_2 = \int_{S(0)} \int_{S(0)} \times \frac{e^{-ik_0(r_0 \cos\alpha + r\cos\beta)}}{\cos\left(\dfrac{\alpha+\theta_0-\alpha_B}{2}\right)\cos\left(\dfrac{\beta+\theta-\alpha_B}{2}\right)\cos\left(\dfrac{\alpha+\beta+\theta_0+\theta}{2}\right)} \times d\alpha\, d\beta. \tag{6.36}$$

It is I_2 alone that contributes the discontinuity across $\theta = \pi - \theta_0$.

From (6.33) and (6.21)

$$L_2(\cos\alpha)\, L_2(-\cos\alpha) = \frac{\mu \sin\alpha + \sqrt{1-\cos^2\alpha/\mu^2}}{\sqrt{1-\cos^2\alpha/\mu^2}} \frac{1}{\sin\alpha + \sin\alpha_B}, \tag{6.37}$$

so that the factor containing the L_2 functions in (6.34) is at least readily evaluated when $\theta = \pi - \theta_0$, which includes the case when transmitter and receiver are at ground level on opposite sides of the boundary line. Aside from this factor, it remains to consider (6.35) and (6.36), and these integrals can be approximated in terms of a comparatively simple single integral, as is shown in the next section.

6.1.4. Reduction of the Solution

To treat the case when transmitter and receiver are on opposite sides of the boundary line it is assumed that θ_0 is close to π and θ close to zero; because of reciprocity the reverse situation does not require separate calculation. Then $|\theta - \alpha_B|$ is small, and (6.36) is approximated by

$$I_2 = \sec\left(\frac{\theta-\alpha_B}{2}\right) \int_{S(0)} \int_{S(0)} \frac{e^{-ik_0(r_0\cos\alpha + r\cos\beta)}}{\cos\left(\dfrac{\alpha+\theta_0-\alpha_B}{2}\right)\cos\left(\dfrac{\alpha+\beta+\theta_0+\theta}{2}\right)} \times d\alpha\, d\beta. \tag{6.38}$$

The change of variables (4.103) gives, approximately,

$$I_2 = \frac{4e^{-ik_0R_1}}{\cos\left(\dfrac{\theta-\alpha_B}{2}\right)\sin\left(\dfrac{\theta_0-\alpha_B}{2}\right)\sin\left(\dfrac{\theta+\theta_0}{2}\right)} \int_{-\infty}^{\infty}\int_{-\infty}^{\infty} \times \frac{e^{-k_0(r_0\xi^2+r\eta^2)}}{\left[\xi - \sqrt{2}\, e^{-\frac{1}{4}i\pi}\cot\left(\dfrac{\theta_0-\alpha_B}{2}\right)\right]\left[\xi+\eta-\sqrt{2}\, e^{-\frac{1}{4}i\pi}\cot\left(\dfrac{\theta+\theta_0}{2}\right)\right]} \times d\xi\, d\eta, \tag{6.39}$$

with $R_1 = r_0 + r$ as before. Certain non-linear terms in ξ, η have been neglected, legitimately since the integrand is only appreciable near $\xi = \eta = 0$.

The further change of variable (4.105) yields

$$I_2 = \frac{4\sqrt{r_0/R_1}\, e^{-ik_0R_1}}{\cos\left(\frac{\theta-\alpha_B}{2}\right)\sin\left(\frac{\theta_0-\alpha_B}{2}\right)\sin\left(\frac{\theta+\theta_0}{2}\right)}\int_0^\infty \varrho K(\varrho)\, e^{-k_0R_1\varrho^2}\, d\varrho, \tag{6.40}$$

where

$$K(\varrho) = \int_0^{2\pi}\left\{\left[\varrho\cos\varphi - e^{-\frac{1}{4}i\pi}\sqrt{\frac{2r_0}{R_1}}\cot\left(\frac{\theta-\alpha_B}{2}\right)\right]\right.$$
$$\times\left.\left[\varrho\left(\sqrt{\frac{r}{R_1}}\cos\varphi + \sqrt{\frac{r_0}{R_1}}\sin\varphi\right) - e^{-\frac{1}{4}i\pi}\frac{\sqrt{2r\,r_0}}{R_1}\cot\left(\frac{\theta+\theta_0}{2}\right)\right]\right\}^{-1}$$
$$\times\, d\varphi. \tag{6.41}$$

Now, albeit with rather more algebra, (6.41) can be evaluated explicitly in precisely the same way as (4.107). When the result is substituted into (6.40) it appears that

$$I_2 = \frac{8\pi i\, e^{-ik_0R_1}}{\cos\left(\frac{\theta-\alpha_B}{2}\right)\sin\left(\frac{\theta_0-\alpha_B}{2}\right)\sin\left(\frac{\theta+\theta_0}{2}\right)}\, I, \tag{6.42}$$

where

$$I = \int_0^\infty \frac{\varrho\, e^{-k_0R_1\varrho^2}\, d\varrho}{\sqrt{\varrho^2-b^2}\left[\sqrt{\varrho^2-b^2} \pm a\sqrt{\frac{R_1}{r_0}} \mp b\sqrt{\frac{r}{r_0}}\right]}$$
$$- \int_0^\infty \frac{\varrho\, e^{-k_0R_1\varrho^2}\, d\varrho}{\sqrt{\varrho^2-a^2}\left[\sqrt{\varrho^2-a^2} + a\sqrt{\frac{r}{r_0}} - b\sqrt{\frac{R_1}{r_0}}\right]}, \tag{6.43}$$

with

$$a = e^{-\frac{1}{4}i\pi}\sqrt{\frac{2r_0}{R_1}}\cot\left(\frac{\theta_0-\alpha_B}{2}\right), \quad b = e^{-\frac{1}{4}i\pi}\frac{\sqrt{2r\,r_0}}{R_1}\cot\left(\frac{\theta+\theta_0}{2}\right), \tag{6.44}$$

and upper sign for $\theta + \theta_0 < \pi$, lower sign for $\theta + \theta_0 > \pi$.

In the first and second integrals of (6.43) make the respective changes of variable $\varrho^2 = \lambda^2 + b^2$, $\varrho^2 = \lambda^2 + a^2$. Then apply to each integral the transformation

$$\int_{i\alpha}^{\infty} \frac{e^{-k_0 R_1 \lambda^2}}{\lambda + \beta}\, d\lambda = e^{-k_0 R_1 \beta^2} \int_{i\sqrt{\alpha^2+\beta^2}}^{\infty} \frac{e^{-k_0 R_1 \lambda^2}}{\lambda}\, d\lambda - \beta \int_{i\alpha}^{\infty} \frac{e^{-k_0 R_1 \lambda^2}}{\lambda^2 - \beta^2}\, d\lambda .$$

The final result is

$$I = \mp \left[a\sqrt{\frac{R_1}{r_0}} - b\sqrt{\frac{r}{r_0}} \right] e^{-k_0 R_1 b^2} \int_{\pm ib}^{\infty} \frac{e^{-k_0 R_1 \lambda^2}\, d\lambda}{\lambda^2 - \left[a\sqrt{\frac{R_1}{r_0}} - b\sqrt{\frac{r}{r_0}} \right]^2}$$

$$+ \left[a\sqrt{\frac{r}{r_0}} - b\sqrt{\frac{R_1}{r_0}} \right] e^{-k_0 R_1 a^2} \int_{ia}^{\infty} \frac{e^{-k_0 R_1 \lambda^2}\, d\lambda}{\lambda^2 - \left[a\sqrt{\frac{r}{r_0}} - b\sqrt{\frac{R_1}{r_0}} \right]^2}, \tag{6.45}$$

with upper sign for $\theta + \theta_0 < \pi$, lower sign for $\theta + \theta_0 > \pi$.

The expression (6.45) gives, with (6.42), the reduced form of I_2, and simpler versions are available only under more special conditions.

The corresponding form for I_1 is easily derived from that for I_2. By taking the approximation analogous to (6.38), and replacing β by $-\beta$, it appears that

$$\cos\left(\frac{\theta - \alpha_B}{2}\right) I_1(\theta) = \cos\left(\frac{\theta + \alpha_B}{2}\right) I_2(-\theta). \tag{6.46}$$

It must be remembered that $I_2(\theta)$ is discontinuous at $\theta = \pi - \theta_0$, but $I_1(\theta)$ is not. Since $\theta_0 - \theta < \pi$, (6.46) holds if $I_2(-\theta)$ is obtained by substituting $-\theta$ for θ in the expression for $I_2(\theta)$ given by the upper sign in (6.45).

To sum up, the diffraction field is given by (6.34), with I_2 given by (6.42), (6.45), and I_1 by (6.46).

Finally, a brief comment is made on the case when source and point of observation are on the same side of the boundary line. In the evaluation of (6.32), it is then permissible to remove the factor

$$\frac{\cos(\frac{1}{2}\alpha)\cos(\frac{1}{2}\beta)}{\cos\alpha + \cos\beta}$$

from under the integral sign at $\alpha = \theta_0$, $\beta = \theta$. For both θ_0 and θ close to π the result is negligible by virtue of the smallness of this factor; whereas for both θ_0 and θ close to zero the remaining non-exponential factors in the integrand can also be replaced by their values at $\alpha = \theta_0$, $\beta = \theta$, and the integral is then easily evaluated explicitly.

6.1.5. Special Cases

Perhaps the most informative way of illustrating the preceding formulae is to consider the case in which the source and point of observation are both in the plane $y = 0$, and take

$$\theta = 0, \quad \theta_0 = \pi, \quad R = S = R_1 = d. \tag{6.47}$$

Then from (6.44)

$$a = e^{-\frac{1}{4}i\pi}\sqrt{2r_0/d}\,\tan(\tfrac{1}{2}\alpha_B), \quad b = 0. \tag{6.48}$$

It is convenient to make use of the parameter

$$\gamma_0 = \sqrt{\tfrac{1}{2}k_0 d}\,\sin\alpha_B \tag{6.49}$$

introduced in (5.51), and write, to an adequate approximation for $|\alpha_B| \ll 1$,

$$a = e^{-\frac{1}{4}i\pi}\sqrt{\frac{r_0}{k_0 d^2}}\,\gamma_0. \tag{6.50}$$

In view of the remarks immediately following (6.46) it is simplest to take the formulae applicable when $\theta + \theta_0$ is (albeit infinitesimally) less than π. Then the asymptotic form of the geometrical optics term (6.31) is

$$H_z^g = \frac{2e^{-ik_0 d}}{\sqrt{k_0 d}}. \tag{6.51}$$

Moreover, (6.37), (6.42) and (6.46) give, respectively,

$$L_2(1)\,L_2(-1) = \frac{1}{\sin\alpha_B}, \quad I_2 = 8\pi i\sec^2(\tfrac{1}{2}\alpha_B)\,e^{-ik_0 d}\,I, \quad I_1 = I_2.$$

Hence the asymptotic form of the diffraction field (6.34) is

$$H_z^d = -4\sqrt{2/\pi}\, e^{-\frac{1}{4}i\pi} \tan(\tfrac{1}{2}\alpha_B)\, e^{-ik_0 d}\, I, \tag{6.52}$$

where I is given by (6.45) with $b = 0$ and the value (6.50) for a. The change of integration variable to $\tau = \sqrt{k_0 d}\,\lambda$ yields

$$I = -e^{-\frac{1}{4}i\pi}\gamma_0 \int_0^\infty \frac{e^{-\tau^2}}{\tau^2 + i\gamma_0^2}\, d\tau + e^{-\frac{1}{4}i\pi}\gamma_0 \sqrt{\frac{r}{d}}\, e^{i\gamma_0^2 r_0/d} \times \int_{e^{\frac{1}{4}i\pi}\gamma_0\sqrt{\frac{r_0}{d}}}^\infty \frac{e^{-\tau^2}}{\tau^2 + i\gamma_0^2 r/d}\, d\tau. \tag{6.53}$$

Now the first term in (6.53) can be replaced by a Fresnel integral (see (3.51)). The combination of (6.51) and (6.52) therefore gives

$$\Delta = 2[1 - 2i\gamma_0 F(\gamma_0)] + \frac{4}{\sqrt{\pi}}\, e^{\frac{1}{4}i\pi}\gamma_0^2 \sqrt{\frac{r}{d}}\, e^{i\gamma_0^2 r/d} \int_{\gamma_0\sqrt{\frac{r_0}{d}}}^\infty \frac{e^{-i\tau^2}}{\tau^2 + \gamma_0^2 r/d}\, d\tau \tag{6.54}$$

as the factor by which the undisturbed source field must be multiplied to obtain the actual field. Note that $\gamma_0\sqrt{r_0/d} = \sqrt{\frac{1}{2}k_0 r_0}\sin\alpha_B$.

If (6.54) is compared with (5.52) it is seen that the difference is in the final term of the former, which is therefore precisely the scattered field generated by the currents in the perfectly conducting half-plane. This term can be further simplified when $|\gamma_0| \gg 1$, since the non-exponential part of the integrand can then be replaced by its value at the lower limit of integration. Thus, for $|\gamma_0| \gg 1$,

$$\Delta = -\frac{4i}{\gamma_0^2} + \frac{4}{\sqrt{\pi}}\, e^{\frac{1}{4}i\pi}\sqrt{\frac{r}{d}}\, F\left[\gamma_0\sqrt{\frac{r_0}{d}}\right]; \tag{6.55}$$

and if also $|\gamma_0\sqrt{r_0/d}| \gg 1$,

$$\Delta = -\frac{4i}{\gamma_0^2} + \frac{2}{\sqrt{\pi}}\, e^{-\frac{1}{4}i\pi}\frac{\sqrt{r/r_0}}{\gamma_0}. \tag{6.56}$$

These formulae predict a recovery of field-strength as r increases from zero, since the second term in (6.56), for example, soon dominates the first. A numerical example is shown in Fig. 6.3.

Curves (a) and (b) are plots of field-strength in decibels against distance in wavelengths from a point-source located respectively above a homogeneous dielectric with $\sin \alpha_B = \frac{1}{3}$ and above an infinite plane perfect conductor. Curve (c) is the corresponding plot for the mixed path, when $k_0 r_0/(2\pi) = 300$; it follows curve (a) up to the boundary, and is subsequently determined by (6.56).

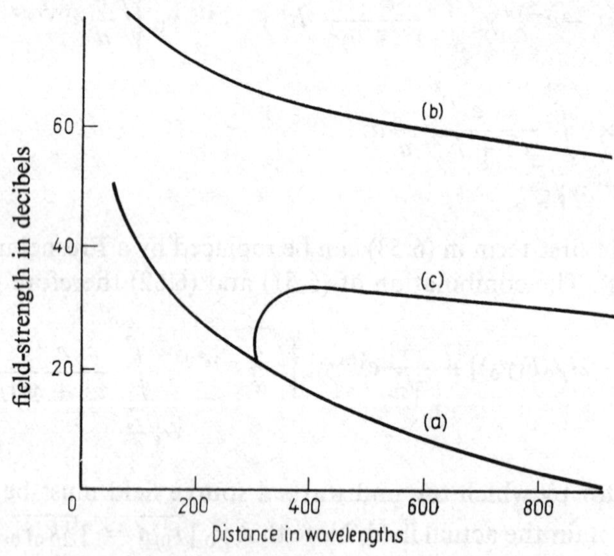

FIG. 6.3.

Finally, it is worth mentioning briefly the other relatively simple approximate solution, corresponding to what was called "ray" theory in § 5.1, and represented there by (5.57). The approximation depends on the heights of source and point of observation being not too small, and consists in putting $\alpha = \theta_0$, $\beta = \theta$ in the arguments of L_1 in the integrand of (6.30). Now if μ, the refractive index of the medium, were unity, L_1 would be unity, and the solution would be just that obtained in § 4.2.5; namely, for $k_0 R \gg 1$, that given by (4.99) and (4.115). Thus in the present case

$$H_z^d = -\frac{\sqrt{2/\pi}\, e^{\frac{1}{4}i\pi}}{L_1(\cos\theta_0)\, L_1(\cos\theta)} \times \left\{\frac{F[\sqrt{k_0(R_1 - R)}]}{\sqrt{k_0(R_1 + R)}} \pm \frac{F[\sqrt{k_0(R_1 - S)}]}{\sqrt{k_0(R_1 + S)}}\right\} e^{-ik_0 R_1}. \tag{6.57}$$

Evidently (6.57) is easily evaluated when $\theta + \theta_0 = \pi$, and in certain circumstances it is possible to verify that the results of this "ray" theory formula agree with those obtained by the application of height-gain functions (see § 5.1.4) to, say, (6.56).

6.2. TWO-PART IMPEDANCE SURFACE

6.2.1. Solution for Incident Plane Wave: *H*-polarization

A problem very similar to that discussed in § 6.1 arises when the surface $y = 0$ can be treated as an impedance surface in the manner of § 5.2, so that the field need be considered only in the region $y \geqq 0$.

The general two-part impedance surface presents no more difficulty than that in which one section is perfectly conducting, and gives rise to a factorization into U and L functions which can be achieved by comparatively elementary methods. The problem can be developed for a line-source just as in §§ 6.1.3–6.1.5, and, of course, has similar application. Here, however, attention is confined to the solution for an incident H-polarized plane wave, with the special object of investigating what happens when a surface wave, supposed launched on a purely reactive surface, encounters a discontinuity in the reactance.

For the purpose of comparison with the preceding analysis the boundary conditions at $y = 0$ are taken as

$$E_x = Z_0 \sin\alpha_1\, H_z, \quad \text{for} \quad x < 0,$$
$$E_x = Z_0 \sin\alpha_2\, H_z, \quad \text{for} \quad x > 0;$$

and the field associated with the incident wave H_z^i, given by (6.1), is conceived as the superposition of H_z^r given by (6.2), where, now,

$$\varrho_H = \frac{\sin\alpha - \sin\alpha_1}{\sin\alpha + \sin\alpha_1}, \tag{6.58}$$

and H_z^s represented by (6.7), with $P(\cos\beta)$ to be found.

The total field is

$$H_z = H_z^i + H_z^r + H_z^s, \tag{6.59}$$

and the boundary conditions at $y = 0$ can be written as

$$E_x^s = Z_0 \sin\alpha_1\, H_z^s, \quad \text{for} \quad x < 0,$$
$$E_x^s = Z_0\{\sin\alpha_2 H_z^s + (\sin\alpha_2 - \sin\alpha_1)(H_z^i + H_z^r)\}, \quad \text{for} \quad x > 0.$$

From (6.7) and (6.8) these lead respectively to the dual integral equations

$$\int_{-\infty}^{\infty}\left[1+\frac{\sin\alpha_1}{\sqrt{1-\lambda^2}}\right]P(\lambda)\,e^{-ik_0x\lambda}\,d\lambda = 0, \quad \text{for} \quad x<0, \tag{6.60}$$

$$\int_{-\infty}^{\infty}\left[1+\frac{\sin\alpha_2}{\sqrt{1-\lambda^2}}\right]P(\lambda)\,e^{-ik_0x\lambda}\,d\lambda = \frac{2\sqrt{1-\lambda_0^2}\,(\sin\alpha_1-\sin\alpha_2)}{\sqrt{1-\lambda_0^2}+\sin\alpha_1}$$
$$\times\, e^{ik_0x\lambda_0}, \quad \text{for} \quad x>0, \tag{6.61}$$

where $\lambda_0 = \cos\alpha$.

Equation (6.60) is satisfied if

$$\left\{1+\frac{\sin\alpha_1}{\sqrt{1-\lambda^2}}\right\}P(\lambda) = U(\lambda), \tag{6.62}$$

and equation (6.61) if

$$\left\{1+\frac{\sin\alpha_2}{\sqrt{1-\lambda^2}}\right\}P(\lambda) = \frac{i}{\pi}\,\frac{\sqrt{1-\lambda_0^2}\,(\sin\alpha_1-\sin\alpha_2)}{\sqrt{1-\lambda_0^2}+\sin\alpha_1}$$
$$\times\,\frac{L(\lambda)}{L(-\lambda_0)(\lambda+\lambda_0)}. \tag{6.63}$$

Write

$$\frac{\sqrt{1-\lambda^2}+\sin\alpha_1}{\sqrt{1-\lambda^2}+\sin\alpha_2} = U_1(\lambda)\,L_1(\lambda), \tag{6.64}$$

where $U_1(\lambda) = L_1(-\lambda)$. Then

$$P(\lambda) = \frac{i}{\pi}\,\frac{\sin\alpha_1-\sin\alpha_2}{[1+\sin\alpha_2/\sqrt{1-\lambda_0^2}][1+\sin\alpha_2/\sqrt{1-\lambda^2}]}$$
$$\times\,\frac{1}{L_1(\lambda_0)\,L_1(\lambda)\,(\lambda+\lambda_0)}, \tag{6.65}$$

and the total field is given by

$$H_z = e^{ik_0r\cos(\theta-\alpha)} + \varrho_H\,e^{-ik_0r\cos(\theta+\alpha)} + H_z^s, \tag{6.66}$$

where

$$H_z^s = \frac{i}{\pi}\,\frac{(\sin\alpha_1-\sin\alpha_2)\sin\alpha}{(\sin\alpha+\sin\alpha_2)\,L_1(\cos\alpha)}$$
$$\times\int_C \frac{\sin\beta\,e^{-ik_0r\cos(\theta-\beta)}}{(\sin\beta+\sin\alpha_2)\,L_1(\cos\beta)\,(\cos\beta+\cos\alpha)}\,d\beta. \tag{6.67}$$

This solution is analogous to (6.24).

Consider now specifically the surface wave problem. For purely reactive surfaces

$$\alpha_1 = i\gamma_1, \quad \alpha_2 = i\gamma_2, \tag{6.68}$$

where γ_1 and γ_2 are taken as real and positive, except in so far as any ambiguity to which this gives rise is resolved by according them small negative imaginary parts. Then with

$$\alpha = \pi - i\gamma_1 \tag{6.69}$$

the incident wave is

$$H_z^i = e^{-k_0 y \sinh\gamma_1} e^{-ik_0 x \cosh\gamma_1}, \tag{6.70}$$

representing a surface wave supported by the surface $y = 0$, $x < 0$ and travelling in the positive x-direction. Also, of course, (6.69) implies $\varrho_H = 0$, so the additional field is simply (6.67), which can now be written

$$H_z^s = -\frac{1}{2\pi}(\sinh\gamma_1 - \sinh\gamma_2)\, L_1(\cosh\gamma_1)$$
$$\times \int_C \frac{\sin\beta\, e^{-ik_0 r\cos(\theta-\beta)}}{(\sin\beta + i\sinh\gamma_2)\, L_1(\cos\beta)\,(\cos\beta - \cosh\gamma_1)}\, d\beta. \tag{6.71}$$

As discussed in § 5.2.2, the resolution of (6.71) into a "space" wave and surface waves is effected by distorting the path of integration into that of steepest descents $S(\theta)$. Now (6.64) is equivalent to

$$\frac{1}{U_1(\cos\beta)\, L_1(\cos\beta)} = \frac{\sin\beta + i\sinh\gamma_2}{\sin\beta + i\sinh\gamma_1}, \tag{6.72}$$

so that $1/L_1(\cos\beta)$ has a pole at $\beta = \pi + i\gamma_1$ and a zero at $\pi + i\gamma_2$ (the pole $-i\gamma_1$ and zero $-i\gamma_2$ of (6.72) belong to $1/U_1(\cos\beta)$). The relevant poles of the integrand of (6.71) are therefore at $-i\gamma_2$,

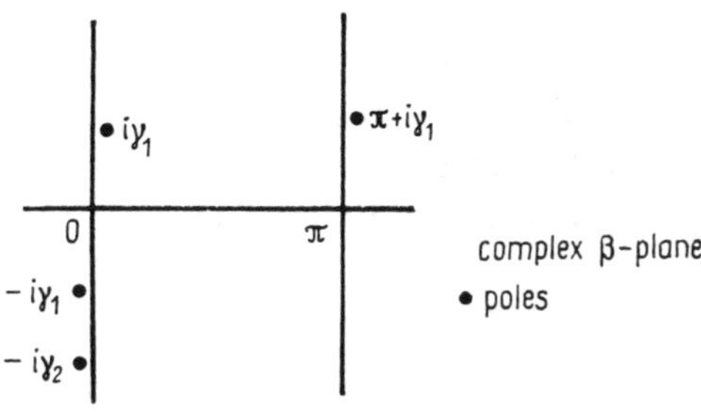

FIG. 6.4.

$\pi + i\gamma_1$ and $\pm i\gamma_1$, as depicted in Fig. 6.4. Evidently, then, in the path distortion, only the poles at $-i\gamma_1$, $-i\gamma_2$ are crossed when θ is close to zero; and only that at $\pi + i\gamma_1$ is crossed when θ is close to π.

The anticipated physical picture is clear: for $\theta \simeq 0$ a transmitted surface wave is contributed by the pole at $-i\gamma_2$, whereas the incident surface wave (6.70) is annulled by the pole at $-i\gamma_1$; for $\theta \simeq \pi$ a reflected surface wave is contributed by the pole at $\pi + i\gamma_1$.

The cancellation of the incident wave, for $\theta \simeq 0$, is easily verified; the pole at $-i\gamma_1$ is encircled anticlockwise, and $2\pi i$ times the product of the residue of the integrand of (6.71) with the factor outside the integral can be written down immediately. It should be remarked that this bit of the analysis does make use of the fact that $L_1(\cos\beta)$ is an even function of β, a feature that has not previously required explicit recognition. As was pointed out at the very end of § 2.2.2, no such implication is to be read into the notation; and evidently, from (6.72), $U_1(\cos\beta)$ cannot also be even in β. The point is explained more thoroughly in the next section, where the "split" in (6.72) is actually effected. It is, of course, necessary to know L_1 in order to obtain the amplitudes and phases of the reflected and transmitted surface waves.

6.2.2. The Split of $\sin\beta + i\sinh\gamma$

The original statement of the required factorization is (6.64), and in the case specified by (6.68) this means, in effect, splitting the function

$$\sqrt{1 - \lambda^2} + i\sinh\gamma, \tag{6.73}$$

where the constant γ has a positive real part and a very small negative imaginary part. The conventional method would be to take the logarithm of the function and express this as the *sum* of functions regular in the upper and lower half-planes respectively, these latter being given by certain contour integrals in the complex λ-plane. For present purposes, however, it is simpler to continue to use the complex β-plane, and to consider the analogous factorization of

$$\sin\beta + i\sinh\gamma; \tag{6.74}$$

it happens that in this way the general contour integration procedure can be avoided.

It is first necessary to appreciate that a λ integral round the branch-cut associated with the branch-point $\lambda = 1$ corresponds to a β integral along the imaginary axis, so that the absence of the branch-point $\lambda = 1$ in a function of λ is equivalent to the corresponding function of β being even in β. Likewise the absence of the branch-point $\lambda = -1$ is equivalent to evenness in $\pi - \beta$. That the L factor of a function of λ has no branch-point at $\lambda = 1$, and the U factor none at $\lambda = -1$, is therefore equivalent to the L factor of the corresponding function of β being an even function of β, and the U factor being an even function of $\pi - \beta$.

Now revert to specific consideration of (6.74). Basically, the aim is to express the logarithm of the function as the sum of appropriate functions. The logarithm itself is not tractable, but can be handled when differentiated, though this, of course, means a subsequent integration. Differentiation with respect to β is a possible technique, but rather neater is differentiation with respect to γ. It is convenient for the subsequent integration to have the logarithm vanish at $\gamma = 0$, so consider

$$f = 1 + i \sinh \gamma / \sin \beta, \tag{6.75}$$

which differs from (6.74) only by the factor $\sin \beta$ with the known split $2 \sin(\frac{1}{2}\beta) \cos(\frac{1}{2}\beta)$.

Since

$$\frac{d}{d\gamma}(\log f) = \frac{i \cosh \gamma}{\sin \beta + i \sinh \gamma} = i \cosh \gamma \frac{\sin \beta - i \sinh \gamma}{\cosh^2 \gamma - \cos^2 \beta}, \tag{6.76}$$

the objective can be achieved by expressing $\sin \beta$ in the numerator as the sum of functions even in $\pi - \beta$ and β, respectively, through the formula

$$\pi \sin \beta = (\pi - \beta) \sin \beta + \beta \sin \beta. \tag{6.77}$$

For in this way (6.76) can be written

$$\frac{d}{d\gamma}(\log f) = g(\pi - \beta, \gamma) + g(\beta, \gamma), \tag{6.78}$$

where

$$g(\beta, \gamma) = \frac{1}{2} \frac{\sinh \gamma}{\cosh \gamma + \cos \beta} + \frac{i}{2\pi} \times \left(\frac{\beta \sin \beta + \gamma \sinh \gamma}{\cosh \gamma - \cos \beta} + \frac{\beta \sin \beta - \gamma \sinh \gamma}{\cosh \gamma + \cos \beta} \right). \tag{6.79}$$

Evidently (6.79) is an even function of β, free of singularities in the semi-infinite strip $0 \leqq \operatorname{Re}\beta \leqq \pi$, $\operatorname{Im}\beta \geqq 0$ (which region maps into the lower half of the complex λ plane). Also it tends to zero as β tends to infinity in the semi-infinite strip, as the form

$$g(\beta,\gamma) = \frac{1}{2}\,\frac{\sinh\gamma}{\cosh\gamma+\cos\beta} + \frac{i}{\pi}\beta\sin\beta\,\frac{\cosh\gamma}{\cosh^2\gamma-\cos^2\beta} + \frac{i}{\pi}\cos\beta\,\frac{\gamma\sinh\gamma}{\cosh^2\gamma-\cos^2\beta} \tag{6.80}$$

makes plain.

Now integrate (6.78) with respect to γ from 0 up to γ, take the exponential, and multiply by $\sin\beta$. This yields the split

$$\sin\beta + i\sinh\gamma = U_2(\cos\beta,\gamma)\,L_2(\cos\beta,\gamma), \tag{6.81}$$

with

$$L_2(\cos\beta,\gamma) = \sqrt{2}\cos(\tfrac{1}{2}\beta)\exp\left\{\int_0^\gamma g(\beta,\tau)\,d\tau\right\}. \tag{6.82}$$

It only remains, therefore, to consider the integration of $g(\beta,\gamma)$.

The first two terms of (6.80) can be integrated explicitly, the third cannot. If β is real, and $0<\beta<\pi$,

$$\int_0^\gamma \frac{\sinh\tau}{\cosh\tau+\cos\beta}\,d\tau = \log\left(\frac{\cosh\gamma+\cos\beta}{1+\cos\beta}\right), \tag{6.83}$$

$$\sin\beta\int_0^\gamma \frac{\cosh\tau}{\cosh^2\tau-\cos^2\beta}\,d\tau = \tan^{-1}\left(\frac{\sinh\gamma}{\sin\beta}\right), \tag{6.84}$$

and

$$L_2(\cos\beta,\gamma) = \sqrt{\cosh\gamma+\cos\beta}\times\exp\left\{\frac{i}{\pi}\beta\tan^{-1}\left(\frac{\sinh\gamma}{\sin\beta}\right)+\frac{i}{\pi}\cos\beta\int_0^\gamma\frac{\tau\sinh\tau}{\cosh^2\tau-\cos^2\beta}\,d\tau\right\}. \tag{6.85}$$

If $\cos\beta$ is real, but $|\cos\beta|>1$, then $\sin\beta$ is purely imaginary and (6.84) is preferably written

$$-\frac{1}{2}i\log\left(\frac{i\sin\beta-\sinh\gamma}{i\sin\beta+\sinh\gamma}\right),$$

and

$$L_2(\cos\beta,\gamma) = \sqrt{\cosh\gamma + \cos\beta}\left(\frac{i\sin\beta - \sinh\gamma}{i\sin\beta + \sinh\gamma}\right)^{\beta/(2\pi)}$$

$$\times \exp\left\{\frac{i}{\pi}\cos\beta\int\limits_0^\gamma \frac{\tau\sinh\tau}{\cosh^2\tau - \cos^2\beta}\,d\tau\right\}. \tag{6.86}$$

Such expressions for $L_2(\cos\beta)$ are quite complicated, but it may be noted that, when $\cos\beta$ is real, $|L_2(\cos\beta)|$ is a simple function. If $\cos\beta$ is real and greater than -1,

$$|L_2(\cos\beta,\gamma)|^2 = \cosh\gamma + \cos\beta. \tag{6.87}$$

This result follows by inspection of (6.85) for $-1 < \cos\beta < 1$ $(0 < \beta < \pi)$ or of (6.86) for $\cos\beta > \cosh\gamma (\beta = -i\beta', \beta' > \gamma)$; and there can be no discontinuity at $\cos\beta = 1$ $(\beta = 0)$ or at $\cos\beta = \cosh\gamma (\beta = -i\gamma)$. If $\cos\beta$ is real and less than -1 the corresponding result can easily be derived from (6.86) by taking $\beta = \pi + i\beta'$, where $\beta' > \gamma$; it is

$$|L_2(\cos\beta,\gamma)|^2 = -(\cosh\gamma + \cos\beta)\frac{i\sin\beta - \sinh\gamma}{i\sin\beta + \sinh\gamma}. \tag{6.88}$$

6.2.3. Surface Wave Reflection and Transmission

The reflected and transmitted surface waves mentioned near the end of § 6.2.1 are now examined more closely.

The magnetic field of the transmitted surface wave is given by $2\pi i$ times the product of the residue of the integrand of (6.71) at the pole $\beta = -i\gamma_2$ with the factor outside the integral. It is

$$\frac{\sinh\gamma_2}{\cosh\gamma_2}\,\frac{\sinh\gamma_1 - \sinh\gamma_2}{\cosh\gamma_1 - \cosh\gamma_2}\,\frac{L_1(\cosh\gamma_1)}{L_1(\cosh\gamma_2)}\,e^{-k_0 y\sinh\gamma_2}\,e^{-ik_0 x\cosh\gamma_2}. \tag{6.89}$$

Note that the incident wave is recovered by letting $\gamma_2 \to \gamma_1$.

The magnetic field of the reflected surface wave is given by $-2\pi i$ times the product of the residue of the integrand of (6.71) at the pole $\beta = \pi + i\gamma_1$ with the factor outside the integral. It is

$$\frac{1}{2}\,\frac{\sinh\gamma_1}{\cosh^2\gamma_1}(\sinh\gamma_1 - \sinh\gamma_2)\,L^2(\cosh\gamma_1)\,e^{-k_0 y\sinh\gamma_1}\,e^{ik_0 x\cosh\gamma_1}. \tag{6.90}$$

In the derivation of (6.90) the residue of $1/L_1(\cos\beta)$ at $\beta = \pi + i\gamma_1$ is obtained by writting (6.72) as

$$\frac{1}{L_1(\cos\beta)} = \frac{\sin\beta + i\sinh\gamma_2}{\sin\beta + i\sinh\gamma_1}U_1(\cos\beta).$$

Now from (6.72) and (6.81),

$$L_1(\cos\beta) = L_2(\cos\beta,\gamma_1)/L_2(\cos\beta,\gamma_2), \tag{6.91}$$

so (6.87) gives

$$\begin{aligned} |L_1(\cosh\gamma_1)|^2 &= \frac{2\cosh\gamma_1}{\cosh\gamma_1 + \cosh\gamma_2}, \\ |L_1(\cosh\gamma_2)|^2 &= \frac{\cosh\gamma_1 + \cosh\gamma_2}{2\cosh\gamma_2}. \end{aligned} \tag{6.92}$$

At least, then, the expressions for the powers carried by the surface waves are simple.

Let p be the ratio of the surface wave power to that carried by the incident wave. Then p for the transmitted surface wave is $\tanh\gamma_1 \coth\gamma_2$ times the square of the modulus of the non-exponential factor in (6.89). Thus, with the help of (6.92),

$$p \text{ transmitted} = \frac{4\sinh\gamma_1\sinh\gamma_2}{(\sinh\gamma_1 + \sinh\gamma_2)^2}. \tag{6.93}$$

Again, p for the reflected surface wave is the square of the modulus of the non-exponential factor in (6.90); hence

$$p \text{ reflected} = \left(\frac{\sinh\gamma_1 - \sinh\gamma_2}{\cosh\gamma_1 + \cosh\gamma_2}\tanh\gamma_1\right)^2. \tag{6.94}$$

The lack of balance in the overall surface wave power is, of course, accounted for by the radiated power associated with the space wave part of (6.71).

CHAPTER VII

THE FIELD OF A MOVING POINT CHARGE

7.1. MOTION IN A PLANE

7.1.1. General Formulation

The derivation of the field due to the motion of individual electrons or protons is a problem of some importance, particularly in the interpretation of various natural processes of radiation. To a first approximation it is usually legitimate to ignore the effect on the particle's motion of the energy lost in radiation, so that the motion can be regarded as specified. The problem is then essentially one of finding the field of a given current distribution, but the solution assumes a characteristic form associated with the fact that the direct information is how the position of a point charge varies with time, rather than how a time-harmonic current density varies in space.

To treat the problem in terms of the plane wave spectrum representation it is assumed that the trajectory is confined to the plane $z = 0$. If the particle has charge e and, at time t, cartesian coordinates

$$[\xi(t),\ \eta(t),\ 0], \tag{7.1}$$

then the suface charge density in $z = 0$ is

$$\sigma = e\delta(x - \xi)\,\delta(y - \eta). \tag{7.2}$$

The corresponding surface current density is therefore

$$\mathbf{j} = e\mathbf{v}\delta(x - \xi)\,\delta(y - \eta), \tag{7.3}$$

where

$$v = (\dot{\xi}, \dot{\eta}, 0) \tag{7.4}$$

is the velocity or the particle, the dot denoting differentiation with respect to t.

The current density must now be expressed in terms of its frequency components. With

$$\mathbf{j}(t) = \int_{-\infty}^{\infty} \mathbf{j}^{\omega}(\omega)\, e^{i\omega t}\, d\omega, \tag{7.5}$$

the frequency spectrum $\mathbf{j}^{\omega}$ is given by

$$\mathbf{j}^{\omega} = \frac{1}{2\pi} \int_{-\infty}^{\infty} \mathbf{j}\, e^{-i\omega t}\, dt. \tag{7.6}$$

It should be noted that, since $\mathbf{j}(t)$ is of necessity real,

$$\mathbf{j}^{\omega *}(\omega) = \mathbf{j}^{\omega}(-\omega); \tag{7.7}$$

thus

$$\int_{-\infty}^{0} \mathbf{j}^{\omega}(\omega)\, e^{i\omega t}\, d\omega = \int_{0}^{\infty} \mathbf{j}^{\omega}(-\omega)\, e^{-i\omega t}\, d\omega = \left[\int_{0}^{\infty} \mathbf{j}^{\omega}(\omega)\, e^{i\omega t}\, d\omega\right]^{*},$$

and (7.5) can be written

$$\mathbf{j} = \operatorname{Re}\ 2\int_{0}^{\infty} \mathbf{j}^{\omega}\, e^{i\omega t}\, d\omega, \tag{7.8}$$

a representation that introduces only positive values of ω.

The substitution of (7.3) into (7.6) gives

$$\mathbf{j}^{\omega} = \frac{e}{2\pi} \int_{-\infty}^{\infty} (\dot{\xi}, \dot{\eta}, 0)\, \delta(x-\xi)\, \delta(y-\eta)\, e^{-i\omega t}\, dt, \tag{7.9}$$

and this result can be used in the formulae presented in § 2.2.4. In particular (2.111) and (2.112) yield

$$[P^{\omega}(l,m), Q^{\omega}(l,m)] = \frac{ek_0^2}{16\pi^3} \iiint_{-\infty}^{\infty} (\dot{\eta}, -\dot{\xi})\, \delta(x-\xi)\, \delta(y-\eta) \times e^{-i\omega t}\, e^{-ik_0(lx+my)}\, dx\, dy\, dt, \tag{7.10}$$

for the plane wave spectrum functions of each frequency component; and if the trivial x- and y-integrations are carried out this leads to

$$(P^{\omega}, Q^{\omega}) = \frac{ek_0^2}{16\pi^3} \int_{-\infty}^{\infty} (\dot{\eta}, -\dot{\xi})\, e^{-ik_0(l\xi+m\eta)}\, e^{-i\omega t}\, dt. \tag{7.11}$$

The frequency spectral resolution of the field vectors is

$$(\mathbf{E}, \mathbf{H}) = \int_{-\infty}^{\infty} (\mathbf{E}^{\omega}, \mathbf{H}^{\omega})\, e^{i\omega t}\, d\omega, \tag{7.12}$$

corresponding to (7.5); and $\mathbf{E}^{\omega}$, $\mathbf{H}^{\omega}$ are given by the appropriate substitution of (7.11) into (2.108), (2.107). Caution is required, however, because the statement in § 2.2.4 that the radical $\sqrt{1 - l^2 - m^2}$ in (2.107), (2.108) is negative pure imaginary when $l^2 + m^2 > 1$ assumes implicitly that ω is positive. If (7.12) is used as it stands, the radical must be taken thus for the part of the range of integration of ω from 0 to ∞, but must be given the reverse sign, when $l^2 + m^2 > 1$, for the remainder of the range of integration of ω from $-\infty$ to 0. Alternatively, (7.12) can be written (cf. (7.8))

$$(\mathbf{E}, \mathbf{H}) = \operatorname{Re} 2 \int_{0}^{\infty} (\mathbf{E}^{\omega}, \mathbf{H}^{\omega})\, e^{i\omega t}\, d\omega, \tag{7.13}$$

and with this prescription the formulae of § 2.2.4 of course require no modification.

One standard expression for the field vectors is obtained by substituting (7.11) into (2.107), (2.108) and then effecting the integration with respect to l and m by means of (2.120). For example, substitution into (2.107) gives

$$\mathbf{H}^{\omega} = \frac{ek_0^2}{16\pi^3} \iiint_{-\infty}^{\infty} \left[\pm\dot{\eta}, \mp\dot{\xi}, \frac{l\dot{\eta} - m\dot{\xi}}{\sqrt{1 - l^2 - m^2}} \right]$$
$$\times\, e^{ik_0[l(x-\xi)+m(y-\eta)\mp\sqrt{1-l^2-m^2}z]}\, e^{-i\omega t}\, dl\, dm\, dt, \tag{7.14}$$

in which the l, m double integrals are seen to be expressible in terms of the respective z, x and y differentials of (2.120). Hence

$$\mathbf{H}^{\omega} = \frac{e}{8\pi^2} \int_{-\infty}^{\infty} [\dot{\eta}z, -\dot{\xi}z, -\dot{\eta}(x - \xi) + \dot{\xi}(y - \eta)] \frac{1}{\varrho}\left(ik_0 + \frac{1}{\varrho}\right)$$
$$\times\, e^{-i\omega(t+\varrho/c)}\, dt, \tag{7.15}$$

where ϱ is the magnitude of

$$\boldsymbol{\varrho} = (x - \xi, y - \eta, z), \tag{7.16}$$

the radius vector to the point of observation from the position of the particle at time t. Evidently (7.16) can be written

$$\mathbf{H}^\omega = \frac{e}{8\pi^2} \int_{-\infty}^{\infty} \left(ik_0 + \frac{1}{\varrho}\right) \frac{\mathbf{v} \wedge \boldsymbol{\varrho}}{\varrho^2} e^{-i\omega(t+\varrho/c)} \, dt, \qquad (7.17)$$

where $\mathbf{v}$ is the velocity of the particl, (7.4). This, in conjunction with (7.12), is the standard form referred to; it is, in fact, also applicable to general three-dimensional motion, and corresponds closely to the expression (3.77) for the magnetic field of a time-harmonic current distribution.

The radiation field of the moving charge can, of course, be written down directly from (7.11), as explained in § 3.2.1. Thus, from (3.22),

$$(H_\theta^\omega, H_\varphi^\omega) \sim \frac{iek_0}{8\pi^2} \frac{e^{-ik_0 r}}{r}$$

$$\times \int_{-\infty}^{\infty} [\dot{\eta} \cos\varphi - \dot{\xi} \sin\varphi, - \cos\theta(\dot{\eta} \sin\varphi + \dot{\xi} \cos\varphi)]$$

$$\times \, e^{ik_0 \sin\theta(\xi\cos\varphi + \eta\sin\varphi)} \, e^{-i\omega t} \, dt, \qquad (7.18)$$

where r, θ, φ are spherical polar coordinates related to x, y, z in the usual way.

If it is asked what power is radiated, it is necessary to be clear about the type of motion postulated. When the time interval during which the particle is in motion, or at least in accelerated motion, is limited, the total energy crossing an elementary area $d\mathbf{S}$ is finite, and of amount $\boldsymbol{W}.d\mathbf{S}$, where

$$\mathbf{W} = \int_{-\infty}^{\infty} \mathbf{E} \wedge \mathbf{H} \, dt, \qquad (7.19)$$

the field vectors being evaluated at $d\mathbf{S}$. In view of (7.12) this gives

$$\mathbf{W} = \iiint_{-\infty}^{\infty} \mathbf{E}^\omega(\omega) \wedge \mathbf{H}^\omega(\omega') \, e^{i(\omega+\omega')t} \, d\omega \, d\omega' \, dt. \qquad (7.20)$$

In (7.20) the t integration introduces $\delta(\omega + \omega')$, and when the ω' integration is then done the result appears as

$$\mathbf{W} = 2\pi \int_{-\infty}^{\infty} \mathbf{E}^\omega(\omega) \wedge \mathbf{H}^\omega(-\omega) \, d\omega. \qquad (7.21)$$

Since (cf. (7.7))

$$\mathbf{E}^{\omega *}(\omega) = \mathbf{E}^{\omega}(-\omega), \quad \mathbf{H}^{\omega *}(\omega) = \mathbf{H}^{\omega}(-\omega), \tag{7.22}$$

(7.21) can be written

$$\mathbf{W} = \operatorname{Re} 4\pi \int_0^\infty \mathbf{E}^{\omega} \wedge \mathbf{H}^{\omega *} \, d\omega . \tag{7.23}$$

The total energy flux density is therefore 8π times the superposition for all positive frequencies of the time-averaged power flux density associated with each frequency component.

A somewhat different situation is that in which the motion of the point charge is periodic. The formulae then appropriate are set out in the next section, and given explicitly in the important case of uniform circular motion.

7.1.2. Periodic Motion: Uniform Circular Motion

Let the motion of the point charge in $z = 0$ be periodic, with period Ω. Then in plane of (7.5), (7.6) the surface current density is written

$$\mathbf{j}(t) = \sum_{-\infty}^{\infty} \mathbf{j}_n e^{in\Omega t} = \mathbf{j}_0 + \operatorname{Re} 2 \sum_{1}^{\infty} \mathbf{j}_n e^{in\Omega t}, \tag{7.24}$$

with

$$\mathbf{j}_n = \mathbf{j}^*_{-n} = \frac{\Omega}{2\pi} \int_0^{2\pi/\Omega} \mathbf{j}(t) \, e^{-in\Omega t} \, dt . \tag{7.25}$$

Ignoring the steady current density $\mathbf{j}_0$, the formula for the harmonic components of the plane wave spectrum functions is evidently

$$(P_n, Q_n) = e \frac{n^2 \Omega^3}{16\pi^3 c^2} \int_0^{2\pi/\Omega} (\dot{\eta}, -\dot{\xi}) \, e^{-in\frac{\Omega}{c}(l\xi + m\eta)} \, e^{-in\Omega t} \, dt . \tag{7.26}$$

The harmonic components of the radiation field are given by

$$(H_{n\theta}, H_{n\varphi}) = \frac{ien\Omega^2}{8\pi^2 c} \frac{e^{-in\Omega r/c}}{r} \times \int_0^{2\pi/\Omega} [\dot{\eta} \cos\varphi - \dot{\xi} \sin\varphi, \; -\cos\theta \, (\dot{\eta} \sin\varphi + \dot{\xi} \cos\varphi)] \times e^{in\frac{\Omega}{c} \sin\theta (\xi \cos\varphi + \eta \sin\varphi)} \, e^{-in\Omega t} \, dt . \tag{7.27}$$

As regards the radiated power, the quantity now of interest is the average over a period $2\pi/\Omega$. If $w.d\mathbf{S}$ is the average rate at which energy crosses an elementary area $d\mathbf{S}$,

$$\mathbf{w} = \frac{\Omega}{2\pi}\int_0^{2\pi/\Omega} \mathbf{E}\wedge\mathbf{H}\,dt. \tag{7.28}$$

In terms of the harmonic components of the field this is

$$\mathbf{w} = \sum_{-\infty}^{\infty} \mathbf{E}_n \wedge \mathbf{H}_n = \mathrm{Re}\, 2\sum_{1}^{\infty} \mathbf{E}_n \wedge \mathbf{H}_n^*, \tag{7.29}$$

again ignoring any steady field components, which are of no significance in this context.

The case in which the point charge travels with constant speed v round a circular orbit is of particular interest, since this is the basic motion induced by an imposed magnetic field. With

$$\xi = \frac{v}{\Omega}\cos(\Omega t), \quad \eta = \frac{v}{\Omega}\sin(\Omega t), \tag{7.30}$$

and φ taken zero without loss of generality, the radiation field (7.27) is seen to be expressible in terms of the integrals

$$\int_0^{2\pi} (\cos\tau, \sin\tau)\, e^{in\frac{v}{c}\sin\theta\cos\tau}\, e^{-in\tau}\, d\tau. \tag{7.31}$$

Now a standard Bessel function representation is

$$J_n(\zeta) = \frac{e^{-\frac{1}{2}in\pi}}{2\pi}\int_0^{2\pi} e^{i(\zeta\cos\tau - n\tau)}\, d\tau, \tag{7.32}$$

so that the integrals in (7.31) are respectively

$$-2\pi i e^{\frac{1}{2}in\pi} J_n'\left(n\frac{v}{c}\sin\theta\right), \quad -\frac{2\pi e^{\frac{1}{2}in\pi}}{\frac{v}{c}\sin\theta} J_n\left(n\frac{v}{c}\sin\theta\right). \tag{7.33}$$

In the radiation field, therefore,

$$H_{n\theta} = \frac{e\Omega}{4\pi} e^{\frac{1}{2}in\pi} n\frac{v}{c} J_n'\left(n\frac{v}{c}\sin\theta\right)\frac{e^{-in\Omega r/c}}{r}, \tag{7.34}$$

$$H_{n\varphi} = -i\frac{e\Omega}{4\pi} e^{\frac{1}{2}in\pi}\, n\cot\theta\, J_n\left(n\frac{v}{c}\sin\theta\right)\frac{e^{-in\Omega r/c}}{r}, \tag{7.35}$$

and the time-averaged power radiated per unit solid angle in the direction θ (for any φ) is

$$Z_0 \frac{e^2\Omega^2}{8\pi^2} \sum_1^\infty n^2 \left[\frac{v^2}{c^2} J_n'^2\left(n\frac{v}{c}\sin\theta\right) + \cot^2\theta\, J_n^2\left(n\frac{v}{c}\sin\theta\right)\right]. \quad (7.36)$$

These formulae can be elucidated with the help of various Bessel function properties. The main point to notice is that if $v/c \ll 1$ the argument of the Bessel functions is much smaller than the order, and the first term of the series (7.36) is predominant. Whereas for a particle of such high energy that $v/c \simeq 1$ argument and order are almost equal in or near the plane of revolution $\theta = \frac{1}{2}\pi$, and then the dominant terms are those for which n is about $\frac{3}{2}(1 - v^2/c^2)^{-3/2}$.

7.2. UNIFORM RECTILINEAR MOTION

7.2.1. Motion in a Vacuum

One of the few cases in which the details of the analysis can be pursued with comparative ease is that in which the point charge travels with constant speed along a straight line. At first sight this is not a problem of much physical interest, but proves otherwise when the charge is allowed to move in a dielectric medium rather than a vacuum.

From the mathematical aspect it is instructive to begin with the vacuum case. Even though its speed is constant, the moving charge constitutes a time-varying current density, and it is pertinent to ask why it does not radiate. From the present point of view the reason is that, at every angular frequency ω, the wavelengths of all the spatial Fourier components of the current density are less than the wavelength $2\pi c/\omega$ of vacuum radiation, and the plane waves of the spectrum representation of the field are therefore all evanescent.

To see this explicitly, let the cartesian coordinates of the point charge e be

$$(vt, 0, 0), \quad (7.37)$$

where v is the constant speed of the charge. In the notation of § 7.1.1,

$$\xi = vt, \quad \eta = 0, \quad (7.38)$$

so (7.9) gives

$$j_x^\omega = \frac{e}{2\pi}\,\delta(y)\,e^{-i\frac{\omega}{v}x}, \quad j_y^\omega = 0. \tag{7.39}$$

It is seen at once that all the spatial wavelengths implicit in (7.39) are less than $2\pi c/\omega$ by virtue of the fact that $v < c$.

The substitution of (7.39) into (2.111) and (2.112) gives

$$P^\omega = 0, \quad Q^\omega = -\frac{ek_0^2}{16\pi^3}\int_{-\infty}^{\infty} e^{-i\frac{\omega}{v}x}\,e^{-ik_0lx}\,dx, \tag{7.40}$$

in agreement, of course, with (7.11). Thus

$$P^\omega = 0, \quad Q^\omega = -\frac{e\omega^2 v}{8\pi^2c^2}\,\delta\left[\omega\left(1+\frac{v}{c}l\right)\right]. \tag{7.41}$$

If now expressions (7.41) are fed into (2.107), the l integration can be done, and it appears that

$$\mathbf{H}^\omega = -\frac{e\omega}{8\pi^2c}\,e^{-i\frac{\omega}{v}x}\int_{-\infty}^{\infty}\left(0, \pm 1, \frac{m}{n_1}\right)e^{i\frac{\omega}{c}(my \mp n_1 z)}\,dm, \tag{7.42}$$

where

$$n_1 = \sqrt{1 - c^2/v^2 - m^2}. \tag{7.43}$$

With $c/v > 1$, n_1 is pure imaginary for all real values of m and the plane waves in the representation (7.42) are all evanescent. The entire field is therefore simply a storage field in the vicinity of the plane $z = 0$, and no energy is lost in radiation.

In a frame of reference moving with the particle there is solely the electrostatic coulomb field of a point charge at rest. A Lorentz transformation will therefore yield an explicit solution to the problem just considered, in terms of simple algebraic functions. How can this solution be derived from (7.42) and (7.13)? The integrals in (7.42) can be expressed in terms of Bessel functions, but these can be avoided by doing first the ω integration of (7.13). The magnetic field lines are clearly circles normal to and centred on the x-axis, so it is sufficient to find $-H_y$ at $y = 0$, $z = \varrho > 0$, since this will give the magnitude of $\mathbf{H}$ at distance ϱ from the line of motion of the charge. It is

$$H = \mathrm{Re}\,\frac{e}{4\pi^2c}\int_{-\infty}^{\infty}\frac{dm}{\left[\frac{\varrho}{c}\sqrt{c^2/v^2 - 1 + m^2} - i(t - x/v)\right]^2}, \tag{7.44}$$

leading to

$$H = \frac{e}{4\pi^2 c} \int_{-\infty}^{\infty} \frac{(\varrho^2/c^2)(c^2/v^2 - 1 + m^2) - (t - x/v)^2}{[(\varrho^2/c^2)(c^2/v^2 - 1 + m^2) + (t - x/v)^2]^2}\, dm. \tag{7.45}$$

The only singularities of the integrand in (7.45) are two poles on the imaginary axis, on either side of the origin, and the integral is easily evaluated by the residue theorem after closing the path of integration with an infinite semicircle. A little algebra yields the standard result

$$H = \frac{e}{4\pi} \frac{v(1 - v^2/c^2)\,\varrho}{[(vt - x)^2 + (1 - v^2/c^2)\,\varrho^2]^{3/2}}. \tag{7.46}$$

It has been shown that in a sense there is no radiation from the charge only because its speed is less than c. This restriction can in effect be overcome in practice by allowing the charge to travel through a dielectric that has refractive index $\mu > c/v$ for at least some range of frequencies. Before considering, in the next section, the analysis for this situation, which gives rise to so-called *Cerenkov radiation,* it is worth examining briefly the hypothetical vacuum case with $v > c$.

With $c/v < 1$, (7.43) is real for some values of m, and these values correspond to homogeneous plane waves which carry energy away to infinity. On the other hand it would seem that there must be a simple explicit solution corresponding to (7.46), and it is instructive to see in what sense this can represent a radiating solution.

The solution can be obtained in the same sort of way as before. The ω integration, with permissible path distortion to ensure convergence, gives

$$H = \text{Re} - \frac{e}{4\pi^2 c} \int_{-\infty}^{\infty} \frac{dm}{\left[\dfrac{\varrho}{c}\sqrt{1 - c^2/v^2 - m^2} - (t - x/v)\right]^2}, \tag{7.47}$$

corresponding to (7.44). Put

$$m = \frac{c\beta}{v}\cos\alpha, \quad \beta = \sqrt{v^2/c^2 - 1}.$$

Then

$$H = \operatorname{Re} - \frac{ev}{4\pi^2 \varrho^2 \beta} \int\limits_C \frac{\sin\alpha \, d\alpha}{\left(\sin\alpha - \dfrac{vt - x}{\beta\varrho}\right)^2}, \tag{7.48}$$

where C is again the path of Fig. 2.1.

It is now observed that it is permissible to distort the path so that it lies entirely along the imaginary axis, allowing only for any poles that may be crossed in the process; and, furthermore, that when the path is along the imaginary axis the real part of the integral is zero, a result easily established by writing $\alpha = i\gamma$ and noting that the real part of the integrand is then an odd function of γ. Thus (7.48) is evaluated merely in terms of the residues of the poles that lie in the semi-infinite strip of the complex α plane $0 < \operatorname{Re}\alpha < \pi$, $\operatorname{Im}\alpha > 0$.

Consider, then the various possible positions of the poles, remembering that $\varrho > 0$. If $vt - x < 0$ the poles are outside the strip $0 < \operatorname{Re}\alpha < \pi$. If $\beta\varrho > vt - x > 0$ there are two poles on the real axis inside the strip; the way in which C is to be indented round these poles is determined by the condition that any imaginary part of $n_1 = (c\beta/v)\sin\alpha$ must be negative, but the details are irrelevant, since the residues of the poles are clearly real and so cannot contribute to (7.48). Finally, if $vt - x > \beta\varrho$ there are poles at

$$\alpha = \tfrac{1}{2}\pi \pm i\alpha_0,$$

where

$$\cosh\alpha_0 = \frac{vt - x}{\beta\varrho};$$

and the residue of the integrand at the pole with positive imaginary part is found to be

$$-\frac{i}{\sinh^3\alpha_0} = -\frac{i\beta^3\varrho^3}{[(vt - x)^2 - \beta^2\varrho^2]^{3/2}}.$$

Thus

$$H = \begin{cases} 0, & \text{for } vt - x < \beta\varrho, \\ -\dfrac{e}{2\pi}\dfrac{v(v^2/c^2 - 1)\,\varrho}{[(vt - x)^2 - (v^2/c^2 - 1)\,\varrho^2]^{3/2}}, & \text{for } vt - x > \beta\varrho. \end{cases} \tag{7.49}$$

The result in this form is a standard one whose broad interpretation is clear. It represents an electromagnetic shock wave, arising from the fact that the particle travels at a speed exceeding that of wave propagation. The shock front is the cone

$$vt - x = \sqrt{v^2/c^2 - 1}\,\varrho, \tag{7.50}$$

which travels with its vertex at the particle. The generators of the cone make the angle $\sin^{-1}(c/v)$ with the particle's track, and the cone can be regarded as the envelope, at any time t, of disturbances emitted by the particle at times prior to t, and subsequently travelling outwards with speed c. The electromagnetic field exists only behind the cone, and becomes infinite as the cone is approached; the factor 2 by which the non-zero part of (7.49) differs in functional form from (7.46) is accounted for by the existence of *two* instants on the particle's track at which emitted disturbances contribute to the field at a given time at a given point behind the cone. For a point in front of the cone, at a given time, there are no such instants.

The singular behaviour of the field accounts for its ability to radiate without conforming to the conventional type of radiation field. But a non-infinite result for the rate of radiation is only obtained with a more realistic model, such as that considered in the next section.

7.2.2. Motion in a Dielectric: Cerenkov Radiation

The present problem is that initially considered in § 7.2.1, except that it is supposed that all space is filled with a homogeneous, lossless, isotropic dielectric of refractive index μ. It is important to recognize that μ is a function of the angular frequency ω; it is taken to tend to unity when ω tends to infinity, as it would on any reasonable model, and the case of interest is that in which $\mu > c/v$ for some, necessarily finite, range of values of ω.

The plane wave spectrum representation for the field in the dielectric medium is only slightly different from (2.107), (2.108). The required modifications are noted at the end of § 2.2.4, but for convenience of reference the explicit statement is now given, namely

$$\mathbf{H} = \iint\limits_{-\infty}^{\infty} \left(\pm P, \pm Q, \frac{lP + mQ}{n} \right) e^{ik_0(lx+my\mp nz)}\, dl\, dm, \tag{7.51}$$

and

$$\mathbf{E} = \frac{Z_0}{\mu^2}$$

$$\times \iint_{-\infty}^{\infty} \left\{ \frac{lmP + (\mu^2 - l^2)Q}{n}, -\frac{(\mu^2 - m^2)P + mlQ}{n}, \mp(mP - lQ) \right\}$$

$$\times\, e^{ik_0(lx + my \mp nz)}\, dl\, dm, \tag{7.52}$$

where

$$n = \sqrt{\mu^2 - l^2 - m^2}, \tag{7.53}$$

and, as usual, the upper/lower sign is for $z \gtrless 0$.

Since (2.111) and (2.112) continue to hold, the frequency components of the spectrum function are given by (7.41) as before. Thus the expression (7.42) for $\mathbf{H}^\omega$ is altered only in that now

$$n_1 = \sqrt{\mu^2 - c^2/v^2 - m^2}. \tag{7.54}$$

The two field components specifically required for the power calculation are

$$H_y^\omega = \mp \frac{e\omega}{8\pi^2 c}\, e^{-i\frac{\omega}{v}x} \int_{-\infty}^{\infty} e^{i\frac{\omega}{c}(my \mp n_1 z)}\, dm, \tag{7.55}$$

$$E_x^\omega = -Z_0 \frac{e\omega}{8\pi^2 c}\left(1 - \frac{c^2}{\mu^2 v^2}\right) e^{-i\frac{\omega}{v}x} \int_{-\infty}^{\infty} \frac{1}{n_1}\, e^{i\frac{\omega}{c}(my \mp n_1 z)}\, dm. \tag{7.56}$$

Consider the total energy that crosses the surface of an infinitely long strip of unit width that lies in some plane $z =$ constant and whose edges are parallel to the y-axis. Allowing for equal radiation on either side of the plane $z = 0$, this quantity is one half the energy, U say, radiated by the point charge per unit length of its path. Thus, using (7.23), and noting that $H_x = 0$,

$$U = \text{Re}\; 8\pi \int_{-\infty}^{\infty}\int_{0}^{\infty} E_x^\omega H_y^{\omega *}\, d\omega\, dy \tag{7.57}$$

evaluated at any value of x and any positive value of z. Of course, U is independent of x and z, which will therefore drop out before the end of the calculations.

The insertion of (7.55) and (7.56) into (7.57) gives

$$U = \operatorname{Re} \frac{Z_0 e^2}{8\pi^3 c^2} \int_0^\infty \left(1 - \frac{c^2}{\mu^2 v^2}\right) \omega^2 \iiint_{-\infty}^{\infty} \frac{1}{n_1} e^{i\frac{\omega}{c}(m-m')y} e^{-i\frac{\omega}{c}(n_1 - n_1'^*)z}$$

$$\times dy\, dm\, dm'\, d\omega,$$

where n_1' is simply the expression (7.54) for n_1 with m replaced by m'. The y integration introduces $\delta\left[\frac{\omega}{c}(m - m')\right]$, so can be followed at once by the m' integration, leaving

$$U = \operatorname{Re} \frac{Z_0 e^2}{4\pi^2 c} \int_0^\infty \left(1 - \frac{c^2}{\mu^2 v^2}\right) \omega \int_{-\infty}^{\infty} \frac{1}{n_1} e^{-i\frac{\omega}{c}(n_1 - n_1^*)z}\, dm\, d\omega. \qquad (7.58)$$

It is now observed that, since μ^2 is real, no contribution is made to (7.58) when n_1 is pure imaginary. Consequently the ω integration can be confined to that range of ω for which $\mu > c/v$, and for any ω in this range the m integration can be confined between the limits $\pm\sqrt{\mu^2 - c^2/v^2}$; this is simply analytic confirmation that radiation occurs only because there are some homogeneous waves in the plane wave spectrum. Moreover, when n_1 is real the exponential factor in the integrand is unity and the m integral is then π. Hence the energy radiated per unit length of path is

$$U = \frac{Z_0 e^2}{4\pi c} \int \left[1 - \frac{c^2}{v^2 \mu^2(\omega)}\right] \omega\, d\omega, \qquad (7.59)$$

taken over all positive values of ω for which $\mu > c/v$. This is the standard formula for Cerenkov radiation.

It may be noted that, since

$$E_x = \operatorname{Re} 2 \int_0^\infty E_x^\omega e^{i\omega t}\, d\omega, \qquad (7.60)$$

with E_x^ω given by (7.56), the formal expression for $-eE_x$ evaluated at the particle ($x = vt$, $y = z = 0$) is the same as (7.59). This is to be expected, in the sense that the work done against the field in maintaining the motion of the particle must balance the rate of loss of energy by radiation. The difficulties inherent in introducing the concept of the field at the location of a point charge are here disguised, in that only the imaginary part of the integral in (7.60) is divergent at $x = vt$, $y = z = 0$.

Consider, finally, what can be said about the directivity of the radiation. Clearly (7.41) states that, for each angular frequency ω for which there is radiation, the plane waves of the spectrum make the specific angle $\cos^{-1}[c/(v\mu)]$ with the particle's track. Thus the directions θ in which there is radiation are those for which

$$\mu(\omega) \cos \theta = c/v \tag{7.61}$$

for some value of ω.

CHAPTER VIII

SOURCES IN ANISOTROPIC MEDIA

8.1. UNIAXIAL MEDIUM

8.1.1. The Dielectric Tensor

This chapter offers some comparatively introductory material on the complicated problems that arise when anisotropic media are involved.

In the notation of § 1.2, the medium considered is supposed to have vacuum permeability and to be characterized, for a time-harmonic field, by a dielectric tensor $\mathscr{K}$ with components $\varkappa_{ij}$ $(i, j = 1, 2, 3)$; moreover the medium is assumed to be lossless, so that $\varkappa_{ji} = \varkappa_{ij}^*$. Two fields of study in which this type of medium is encountered are crystal optics and magneto-ionic theory; in the latter the medium is a gas of free electrons (with a "neutralizing background" of positive ions, often supposed immobile) permeated by an imposed magnetostatic field.

In crystal optics the simplest case of anisotropy is afforded by the uniaxial crystal for which, with appropriate choice of coordinate axes,

$$\mathscr{K} = \begin{pmatrix} \varkappa & 0 & 0 \\ 0 & \varkappa & 0 \\ 0 & 0 & \varkappa' \end{pmatrix}, \tag{8.1}$$

and $\varkappa$, $\varkappa'$ are real and positive. This form also arises in magneto-ionic theory when the magnetostatic field B_0 is so strong that the angular frequency of free gyration of an electron in the field (eB_0/m) is very high compared with both the wave angular frequency ω and the so-called plasma frequency ω_p, the latter being given by

$$\omega_p^2 = Ne^2/(\varepsilon_0 m), \tag{8.2}$$

where $-e$, m are the electron charge and mass, and N is the number of electrons per unit volume. Then, with the z-axis directed along $\mathbf{B}_0$, the dielectric tensor is, on the simplest theory, approximately

$$\mathscr{K} = \begin{pmatrix} 1 & 0 & 0 \\ 0 & 1 & 0 \\ 0 & 0 & 1 - \omega_p^2/\omega^2 \end{pmatrix}. \tag{8.3}$$

It is easy to see how the form (8.3) arises. If $\mathbf{E}$ is in the z-direction, that is, along $\mathbf{B}_0$, the electrons tend to oscillate parallel to that direction, and such motion is unaffected by $\mathbf{B}_0$; the equivalent dielectric constant is therefore the same as that of a free electron gas without a magnetostatic field, and since on the simplest theory the current density is $\mathbf{J} = -Ne\mathbf{v} = Ne^2\mathbf{E}/(i\omega m)$, this latter is $1 - \omega_p^2/\omega^2$. If, on the other hand, $\mathbf{E}$ is transverse to the z-direction, the electrons tend to oscillate only in planes normal to $\mathbf{B}_0$; but then the amplitudes of the oscillations are negligible owing to the high value of $\mathbf{B}_0$. The equivalent dielectric constant is therefore the same as that in the absence of the electrons, namely unity.

Formally, of course, (8.3) is obtained from (8.1) on writing

$$\varkappa = 1, \quad \varkappa' = 1 - \omega_p^2/\omega^2, \tag{8.4}$$

but in this identification it is necessary to admit the possibility that $\varkappa'$ may be negative; this case arises when $\omega < \omega_p$, and is of interest as a model which can be used to illustrate the generation of Cerenkov radiation in a magneto-ionic medium, in the way shown in § 8.1.6.

8.1.2. Surface Currents in Plane Normal to Axis

Consider the field, in a uniaxial medium, generated by a distribution of surface current density in a plane that is normal to the direction of symmetry of the medium. The dielectric tensor of the medium is supposed given by (8.1), and $z = 0$ is taken as the current carrying plane.

The intention is to apply the plane wave spectrum representation introduced in § 2.2.6. To this end it is necessary to confirm that the symmetry about $z = 0$ assumed in (2.124) holds for the present case, and to find n_1, n_2, P_1/Q_1, P_2/Q_2 as functions of l and m.

From a trivial generalization of the analysis given in § 2.1.4, plane waves with space factor

$$e^{ik_0(lx+my\mp nz)} \tag{8.5}$$

can be supported by the medium under one or other of the conditions (cf. (2.42), (2.43))

$$l^2 + m^2 + n^2 = \varkappa, \tag{8.6}$$

$$l^2 + m^2 + (\varkappa'/\varkappa)\, n^2 = \varkappa'. \tag{8.7}$$

Thus n^2 as a function of l and m must take one of the two forms

$$n_1^2 = \varkappa - l^2 - m^2, \tag{8.8}$$

$$n_2^2 = \varkappa - (\varkappa/\varkappa')\,(l^2 + m^2). \tag{8.9}$$

The corresponding refractive indices are

$$\mu_1^2 = \varkappa, \tag{8.10}$$

$$\mu_2^2 = \varkappa + (1 - \varkappa/\varkappa')\,(l^2 + m^2). \tag{8.11}$$

To establish the validity of (2.124) consider first the general expression

$$\mathbf{H} = \sum_{i=1,2} \left(P_i^\pm, Q_i^\pm, \pm \frac{lP_i^\pm + mQ_i^\pm}{n_i}\right) e^{ik_0(lx+my\mp n_i z)} \tag{8.12}$$

for the magnetic field of the pair of plane waves emitted into the respective half-spaces $z > 0$ (upper sign), $z < 0$ (lower sign) by a space-harmonic surface current density in the plane $z = 0$. Then the components of

$$\operatorname{curl} \mathbf{H} = i\omega\varepsilon_0 \mathscr{K} \mathbf{E} = i\omega\varepsilon_0(\varkappa E_x, \varkappa E_y, \varkappa' E_z)$$

yield the associated electric field

$$\mathbf{E} = Z_0 \sum_{i=1,2} \left[\pm \frac{lmP_i^\pm + (\mu_i^2 - l^2)\, Q_i^\pm}{\varkappa n_i}, \mp \frac{(\mu_i^2 - m^2)\, P_i^\pm + mlQ_i^\pm}{\varkappa n_i}, - \frac{mP_i^\pm - lQ_i^\pm}{\varkappa'}\right] e^{ik_0(ln+my\mp n_i z)}. \tag{8.13}$$

Furthermore, as explained, in effect, in § 2.1.4, a plane wave with refractive index μ_1 has the structure of a plane wave in an isotropic dielectric of dielectric constant $\varkappa$, and its polarization is such that $E_z = 0$ (eqn. (2.44)); hence

$$mP_1^\pm - lQ_1^\pm = 0. \tag{8.14}$$

Again, it is easy to verify, for example from (2.37), that a plane wave with refractive index μ_2 has polarization such that $H_z = 0$; hence

$$lP_2^{\pm} + mQ_2^{\pm} = 0. \tag{8.15}$$

But E_x and E_y must be continuous across $z = 0$, and this evidently demands

$$P_i^- = -P_i^+, \quad Q_i^- = -Q_i^+, \quad (i = 1, 2). \tag{8.16}$$

The representation (2.124) is therefore vindicated, with (8.8), (8.9), (8.14), (8.15) giving the information required for its application.

8.1.3. Dipole Normal to Axis

A simple illustration of the analysis of § 8.1.2 is afforded by the electric dipole. If this is located at the origin and directed along the x axis, say, the surface current density components are, as in (2.114),

$$j_x = p\,\delta(x)\,\delta(y), \quad j_y = 0. \tag{8.17}$$

Then, from (2.126),

$$P_1 + P_2 = 0, \quad Q_1 + Q_2 = -\frac{pk_0^2}{8\pi^2}, \tag{8.18}$$

analogous, of course, to (2.115); and (8.14), (8.15), (8.18) give

$$(P_1, Q_1, P_2, Q_2) = -\frac{pk_0^2}{8\pi^2}\frac{1}{l^2 + m^2}(lm, m^2, -lm, l^2). \tag{8.19}$$

The representation (2.123), (2.124) is therefore

$$\mathbf{H} = \mathbf{H}_1 + \mathbf{H}_2, \tag{8.20}$$

where

$$\mathbf{H}_1 = -\frac{pk_0^2}{8\pi^2}\iint_{-\infty}^{\infty}\left(\pm\frac{lm}{l^2 + m^2}, \pm\frac{m^2}{l^2 + m^2}, \frac{m}{n_1}\right) \times e^{ik_0(lx+my\mp n_1 z)}\,dl\,dm, \tag{8.21}$$

$$\mathbf{H}_2 = -\frac{pk_0^2}{8\pi^2}\iint_{-\infty}^{\infty}\left(\mp\frac{lm}{l^2 + m^2}, \pm\frac{l^2}{l^2 + m^2}, 0\right) e^{ik_0(lx+my\mp n_2 z)}\,dl\,dm, \tag{8.22}$$

with n_1, n_2 given by (8.8) and (8.9).

Evidently

$$\mathbf{H}_1 = \left(\frac{\partial^2 M}{\partial x\, \partial z}, \frac{\partial^2 M}{\partial y\, \partial z}, -\frac{\partial^2 M}{\partial x^2} - \frac{\partial^2 M}{\partial y^2}\right), \tag{8.23}$$

$$\mathbf{H}_2 = ik_0 Y_0 \left(\frac{\partial \Pi}{\partial y}, -\frac{\partial \Pi}{\partial x}, 0\right), \tag{8.24}$$

where

$$M = -\frac{p}{8\pi^2} \iint\limits_{-\infty}^{\infty} \frac{m}{n_1(l^2+m^2)} e^{ik_0(lx+my\mp n_1 z)}\, dl\, dm, \tag{8.25}$$

$$\Pi = \mp \frac{pZ_0}{8\pi^2} \iint\limits_{-\infty}^{\infty} \frac{l}{l^2+m^2} e^{ik_0(lx+my\mp n_2 z)}\, dl\, dm. \tag{8.26}$$

Reference to (8.8), (8.9) shows that, if for the present both $\varkappa$ and $\varkappa'$ are assumed positive, (8.25), (8.26) can be written

$$M = -\frac{p}{8\pi^2} \iint\limits_{-\infty}^{\infty} \frac{m}{n(l^2+m^2)} e^{ik_0\sqrt{\varkappa}(lx+my\mp nz)}\, dl\, dm, \tag{8.27}$$

$$\Pi = \mp \frac{pZ_0}{8\pi^2} \sqrt{\varkappa'} \iint\limits_{-\infty}^{\infty} \frac{l}{l^2+m^2} e^{ik_0(l\sqrt{\varkappa'}\,x+m\sqrt{\varkappa'}\,y\mp n\sqrt{\varkappa}\,z)}\, dl\, dm, \tag{8.28}$$

where

$$n = \sqrt{1-l^2-m^2}. \tag{8.29}$$

Hence it is only necessary to evaluate

$$I(x, y, z) = \iint\limits_{-\infty}^{\infty} \frac{l}{n(l^2+m^2)} e^{ik_0(lx+my\mp nz)}\, dl\, dm, \tag{8.30}$$

since

$$M = -\frac{p}{8\pi^2} I(\sqrt{\varkappa}\,y, \sqrt{\varkappa}\,x, \sqrt{\varkappa}\,z), \tag{8.31}$$

$$\Pi = -\frac{ipZ_0}{8\pi^2 k_0} \sqrt{\frac{\varkappa'}{\varkappa}} \frac{\partial}{\partial z} I(\sqrt{\varkappa'}\,x, \sqrt{\varkappa'}\,y, \sqrt{\varkappa}\,z). \tag{8.32}$$

The evaluation of I can be effected by much the same analysis as that used at the end of § 2.2.5. The transformations leading from

(2.120) to (2.121) give, for $z > 0$,

$$I = 2\pi i \cos\chi \int_0^\infty J_1(k_0\varrho\tau) \frac{e^{-ik_0 z\sqrt{1-\tau^2}}}{\sqrt{1-\tau^2}}\, d\tau, \tag{8.33}$$

and this integral can in turn be derived from (2.121) as follows. If the familiar relation

$$\frac{d}{d\eta} J_1(\eta) + \frac{1}{\eta} J_1(\eta) = J_0(\eta)$$

is multiplied by η and then integrated over η from 0 to η_0 there results

$$\int_0^{\eta_0} \eta\, J_0(\eta)\, d\eta = \eta_0 J_1(\eta_0),$$

from which

$$\tau \int_0^{k_0\varrho} \xi J_0(\xi\tau)\, d\xi = k_0\varrho J_1(k_0\varrho\tau).$$

Now, in both sides of (2.121), first replace $k_0\varrho$ by ξ, next multiply by ξ, and then integrate over ξ from 0 to $k_0\varrho$ to get

$$k_0\varrho \int_0^\infty J_1(k_0\varrho\tau) \frac{e^{-ik_0 z\sqrt{1-\tau^2}}}{\sqrt{1-\tau^2}}\, d\tau = e^{-ik_0 z} - e^{-ik_0\sqrt{\varrho^2+z^2}}. \tag{8.34}$$

Hence (8.33) is

$$I = 2\pi i \cos\chi \frac{e^{-ik_0 z} - e^{-ik_0 r}}{k_0\varrho}, \tag{8.35}$$

where

$$x = \varrho\cos\chi, \quad y = \varrho\sin\chi, \quad r^2 = x^2 + y^2 + z^2. \tag{8.36}$$

The substitution of (8.35) into (8.31), (8.32) gives

$$M = -\frac{ip}{4\pi k_0\sqrt{\varkappa}} \frac{y}{x^2+y^2} \left(e^{-ik_0\sqrt{\varkappa}z} - e^{-ik_0\sqrt{\varkappa}r}\right), \tag{8.37}$$

$$ik_0 Y_0 \Pi = \frac{p}{4\pi} \frac{x}{x^2+y^2} \left(e^{-ik_0\sqrt{\varkappa}z} - \sqrt{\varkappa}\frac{z}{R} e^{-ik_0 R}\right), \tag{8.38}$$

where

$$R^2 = \varkappa'(x^2+y^2) + \varkappa z^2. \tag{8.39}$$

To get the components of **H** it only remains to carry out the differentiation specified in (8.23), (8.24). The result is

$$H_x = \frac{pk_0^2}{4\pi}\sqrt{\varkappa}\,\frac{k_0 xyz}{x^2+y^2}\left[\frac{2}{k_0^2(x^2+y^2)}\left(\frac{e^{-ik_0R}}{k_0R}-\frac{e^{-ik_0\sqrt{\varkappa}r}}{k_0\sqrt{\varkappa}\,r}\right)\right.$$

$$\left.-\frac{i}{k_0^2r^2}\left(1-\frac{i}{k_0\sqrt{\varkappa}r}\right)e^{-ik_0\sqrt{\varkappa}r}+\frac{i\varkappa'}{k_0^2R^2}\left(1-\frac{i}{k_0R}\right)e^{-ik_0R}\right], \tag{8.40}$$

$$H_y = -\frac{pk_0^2}{4\pi}\sqrt{\varkappa}\,\frac{z}{k_0(x^2+y^2)}\left[\frac{x^2-y^2}{x^2+y^2}\left(\frac{e^{-ik_0R}}{k_0R}-\frac{e^{-ik_0\sqrt{\varkappa}r}}{k_0\sqrt{\varkappa}\,r}\right)\right.$$

$$\left.+i\frac{y^2}{r^2}\left(1-\frac{i}{k_0\sqrt{\varkappa}r}\right)e^{-ik_0\sqrt{\varkappa}r}+i\varkappa'\frac{x^2}{R^2}\left(1-\frac{i}{k_0R}\right)e^{-ik_0R}\right], \tag{8.41}$$

$$H_z = \frac{pk^2}{4\pi}i\sqrt{\varkappa}\,\frac{y}{k_0r^2}\left(1-\frac{i}{k_0\sqrt{\varkappa}r}\right)e^{-ik_0\sqrt{\varkappa}r}. \tag{8.42}$$

Two points may be noted. First, that the expression $\exp(-ik_0\sqrt{\varkappa}z)$ only makes an interim appearance, in (8.37) and (8.38). When M and Π are combined to form the field components the terms involving this expression cancel one another, as would be anticipated, and there is ultimately no restriction on the sign of z. Secondly, that although the final formulae were derived on the assumption that $\varkappa' > 0$, they must also apply to the physically attainable case $\varkappa' < 0$ (noted at the end of § 8.1.1), provided that R is taken to be negative pure imaginary for those field-points for which (8.39) is negative. In this case, then, the factor $\exp(-ik_0R)$ characterizes either a propagated or an evanescent field according to whether the field-point lies inside or outside the double conical surface whose generators pass through the origin and make the angle

$$\tan^{-1}\sqrt{|\varkappa/\varkappa'|} \tag{8.43}$$

with the z-axis. The field components become indefinitely large as the surface of the cone is approached.

In spherical polar coordinates r, θ, φ, with the dipole as origin, direction of dipole as axis ($x = r\cos\theta$), and φ measured from the xz-plane it is found that

$$H_r = \frac{pk_0^2}{4\pi}\sqrt{\varkappa}\,\frac{\sin^2\theta\sin\varphi\cos\varphi}{1-\sin^2\theta\cos^2\varphi}\,\frac{1}{k_0r}\left(\frac{e^{-ik_0\sqrt{\varkappa}r}}{k_0\sqrt{\varkappa}\,r}-\frac{e^{-ik_0R}}{k_0R}\right), \tag{8.44}$$

and

$$H_\theta = \frac{pk_0^2}{4\pi} i \sqrt{\varkappa} \frac{\sin\theta\cos\theta\sin\varphi\cos\varphi}{1-\sin^2\theta\cos^2\varphi} \left[\frac{1+\sin^2\theta\cos^2\varphi}{1-\sin^2\theta\cos^2\varphi} \frac{i}{k_0 r} \right.$$

$$\times \left(\frac{e^{-ik_0\sqrt{\varkappa}\,r}}{k_0\sqrt{\varkappa}\,r} - \frac{e^{-ik_0 R}}{k_0 R} \right) - \frac{1}{k_0 r}\left(1 - \frac{i}{k_0\sqrt{\varkappa}\,r}\right) e^{-ik_0\sqrt{\varkappa}\,r}$$

$$\left. + \varkappa' \frac{r}{k_0 R^2}\left(1 - \frac{i}{k_0 R}\right) e^{-ik_0 R} \right], \tag{8.45}$$

$$H_\varphi = \frac{pk_0^2}{4\pi} i \sqrt{\varkappa} \frac{\sin\theta}{\tan^2\varphi + \cos^2\theta} \left[\frac{\cot^2\theta - \sin^2\varphi}{\cot^2\theta + \sin^2\varphi} \frac{i}{k_0 r} \right.$$

$$\times \left(\frac{e^{-ik_0\sqrt{\varkappa}\,r}}{k_0\sqrt{\varkappa}\,r} - \frac{e^{-ik_0 R}}{k_0 R} \right) + \frac{\tan^2\varphi}{k_0 r}\left(1 - \frac{i}{k_0\sqrt{\varkappa}\,r}\right) e^{-ik_0\sqrt{\varkappa}\,r}$$

$$\left. + \varkappa' \cos^2\theta \frac{r}{k_0 R^2}\left(1 - \frac{i}{k_0 R}\right) e^{-ik_0 R} \right]. \tag{8.46}$$

When $\varkappa' = \varkappa$, (8.44) and (8.45) vanish, and the familiar form of the field of a dipole in an isotropic dielectric is recovered.

8.1.4. Surface Currents in Plane Parallel to Axis

Here the case is considered in which the plane of the surface current distribution is parallel to the direction of symmetry of the uniaxial medium. It is convenient to retain $z = 0$ as the current carrying plane, and the dielectric tensor of the medium is consequently taken to be

$$\mathscr{K} = \begin{pmatrix} \varkappa' & 0 & 0 \\ 0 & \varkappa & 0 \\ 0 & 0 & \varkappa \end{pmatrix}. \tag{8.47}$$

The validity of the representation (2.123), (2.124) can be established just as in § 8.1.2, with now

$$n_1^2 = \varkappa - l^2 - m^2, \tag{8.48}$$

$$n_2^2 = \varkappa' - (\varkappa'/\varkappa)\, l^2 - m^2. \tag{8.49}$$

The plane wave characterized by (8.48) has $E_x = 0$, so that

$$lmP_1 + (\varkappa - l^2)\, Q_1 = 0; \tag{8.50}$$

and that characterized by (8.49) has $H_x = 0$, so that

$$P_2 = 0. \tag{8.51}$$

8.1.5. Dipole Parallel to Axis

When the surface current density is (8.17), specifying a dipole parallel to the direction of symmetry of the medium whose dielectric tensor is (8.47), the results are particularly simple. They can, of course, be used to supplement those of § 8.1.3 to yield, by superposition, the solution for an arbitrarily oriented dipole.

Equations (8.18), (8.50) and (8.51) imply

$$P_1 = P_2 = Q_1 = 0, \quad Q_2 = -\frac{pk_0^2}{8\pi^2}. \tag{8.52}$$

Hence, from (2.124), $\mathbf{H}_1 = 0$, and the components of $\mathbf{H}_2$ can be written

$$H_{2x} = 0, \quad H_{2y} = ik_0 Y_0 \frac{\partial \Pi}{\partial z}, \quad H_{2z} = -ik_0 Y_0 \frac{\partial \Pi}{\partial y}, \tag{8.53}$$

where

$$\Pi = -\frac{pZ_0}{8\pi^2} \iint_{-\infty}^{\infty} \frac{1}{n_2} e^{ik_0(lx+my\mp n_2 z)}\, dl\, dm. \tag{8.54}$$

Since n_2 is given by (8.49) it is evident that

$$\Pi = -\frac{ipZ_0}{4\pi} \sqrt{\varkappa}\, \frac{e^{-ik_0 R}}{k_0 R}, \tag{8.55}$$

where

$$R^2 = \varkappa x^2 + \varkappa'(y^2 + z^2). \tag{8.56}$$

The physical variables (8.56) and (8.39) are identical when the present dipole and that in § 8.1.3 are placed orthogonally to one another at the same point in a common medium.

The cartesian components of $\mathbf{H}$ are obtained by differentiation of (8.55) as indicated in (8.53), and those of $\mathbf{E}$ can then be derived by forming curl $\mathbf{H}$. If spherical polar cordinates r, θ, φ are introduced, with the dipole as origin and direction of dipole as axis ($x = r\cos\theta$), it is found that the non-zero components are

$$H_\varphi = \frac{pk_0^2}{4\pi} i\varkappa' \sqrt{\varkappa} \sin\theta \frac{r}{R}\left(1 - \frac{i}{k_0 R}\right)\frac{e^{-ik_0 R}}{k_0 R}, \tag{8.57}$$

$$E_\theta = Z_0 \frac{pk_0^2}{4\pi} i\varkappa' \sqrt{\varkappa} \sin\theta \frac{r^2}{R^2}$$
$$\times \left\{1 - \left[1 + 2\left(1 - \frac{\varkappa}{\varkappa'}\right)\cos^2\theta\right]\frac{i}{k_0 R}\left(1 - \frac{i}{k_0 R}\right)\right\}\frac{e^{-ik_0 R}}{k_0 R}, \tag{8.58}$$

$$E_r = Z_0 \frac{pk_0^2}{4\pi} 2\sqrt{\varkappa} \cos\theta \frac{1}{k_0 R}\left(1 - \frac{i}{k_0 R}\right)\frac{e^{-ik_0 R}}{k_0 R}, \tag{8.59}$$

where

$$R^2 = (\varkappa \cos^2 \theta + \varkappa' \sin^2 \theta) r^2. \tag{8.60}$$

When $\varkappa' = \varkappa$ the familiar form of the field of a dipole in an isotropic dielectric is recovered.

If $\varkappa' < 0$ the field is propagated or evanescent, respectively, in the regions inside or outside the conical surface

$$\tan^2 \theta = |\varkappa/\varkappa'|, \tag{8.61}$$

and the field components become indefinitely large as the surface of the cone is approached. This behaviour is, of course, conditioned by the properties of the medium in much the same way as that of the field described in § 8.1.3.

8.1.6. Point Charge in Uniform Motion Parallel to Axis

A second simple illustration of the analysis of § 8.1.4 is the case in which a point charge travels with constant speed v along a line parallel to the direction of symmetry of the medium. If the medium is an electron gas in a strong magnetostatic field,

$$\varkappa = 1, \quad \varkappa' = 1 - \omega_p^2/\omega^2,$$

as in (8.4), and the occurrence of Cerenkov radiation may be anticipated for the following reason. The plane wave characterized by (8.49) has refractive index μ_2, where (cf. (2.47))

$$\mu_2^2 = \frac{\varkappa'}{\sin^2 \theta + \varkappa' \cos^2 \theta}, \tag{8.62}$$

θ being the angle which the direction of phase propagation makes with the magnetostatic field. The (plane) radial plot of μ_2 against θ, sketched in Fig. 8.1, is an ellipse when $\varkappa' > 0$ $(\omega > \omega_p)$ and a hyperbola when $\varkappa' < 0$ $(\omega < \omega_p)$. Thus the condition (7.61) for Cerenkov radiation,

$$\mu_2 \cos \theta = c/v,$$

can be satisfied at angular frequency ω for every value of ω less than ω_p, but for no value greater than ω_p.

Figure 8.1 also provides important information about the different directions of phase and energy propagation. For a direction of phase propagation θ, represented by the point P on the $\mu_2(\theta)$ curve, the corresponding direction of energy propagation is known to be along the normal to the curve at P. When $\varkappa' < 0$ the z-components of these directions are therefore of opposite sign, a fact which is confirmed by the finding that if the direction of phase propagation is $(-l, -m, n_2)$ that of $\mathbf{E} \wedge \mathbf{H}^*$ is $(\varkappa' l, m, -n_2)$. This means that in the plane wave spectrum representation (2.124) the real values of n_2 must be taken to be *negative*.

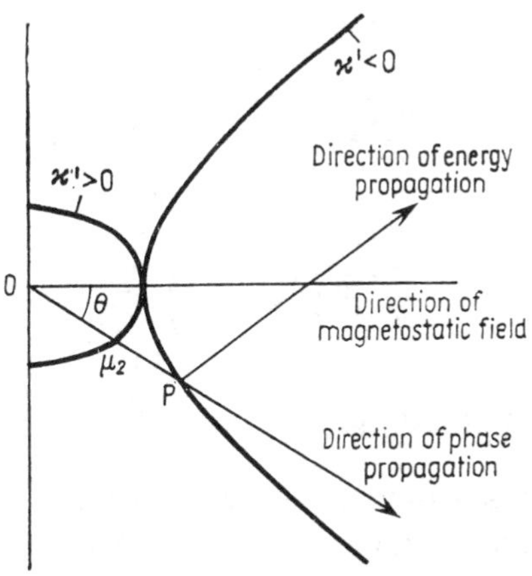

FIG. 8.1.

To obtain the frequency components of the plane wave spectrum functions it is only necessary to combine the appropriate version of (7.41), that is

$$P_1^\omega + P_2^\omega = 0, \quad Q_1^\omega + Q_2^\omega = -\frac{e\omega}{8\pi^2 c}\,\delta\left(l + \frac{c}{v}\right), \tag{8.63}$$

with (8.50) and (8.51); the result is

$$P_1^\omega = P_2^\omega = Q_1^\omega = 0, \quad Q_2^\omega = -\frac{e\omega}{8\pi^2 c}\,\delta\left(l + \frac{c}{v}\right). \tag{8.64}$$

The representation (2.124) therefore gives

$$\mathbf{H}^\omega = -\frac{e\omega}{8\pi^2 c}\,e^{-i\omega x/v}\int_{-\infty}^{\infty}\left(0, \pm 1, \frac{m}{n_2'}\right) e^{i\frac{\omega}{c}(my \mp n_2' z)}\,dm; \tag{8.65}$$

here

$$n_2' = \sqrt{\varkappa'(1 - c^2/v^2) - m^2}\,, \tag{8.66}$$

and, for real m, is either negative real or negative pure imaginary.

As explained at the end of § 7.2.2 the energy U radiated per unit length of the particle's track can be found most quickly by identifying it with $-eE_x$ evaluated at the particle. Since, corresponding to (8.65),

$$E_x^\omega = \frac{eZ_0}{8\pi^2 c}\left(\frac{c^2}{v^2} - 1\right)\omega\, e^{-i\omega x/v} \int_{-\infty}^{\infty} \frac{1}{n_2'}\, e^{i\frac{\omega}{c}(my \mp n_2' z)}\, dm, \tag{8.67}$$

it follows that

$$U = \frac{e^2 Z_0}{4\pi c}\left(\frac{c^2}{v^2} - 1\right)\int_0^{\omega_p} \omega\, d\omega;$$

that is

$$U = \frac{e^2 Z_0}{8\pi c}\left(\frac{c^2}{v^2} - 1\right)\omega_p^2. \tag{8.68}$$

8.1.7. TE and TM Resolution

In this section it is shown how the problem of the field of sources in a uniaxial medium can be expressed in terms of a corresponding vacuum field problem.

Attention is drawn to two features of the preceding discussion; first, the nature of the polarization of the plane waves supported by the uniaxial medium, and secondly, the coordinate scaling implicit, for example, in the arguments of I in (8.31) and (8.32).

For a medium with dielectric tensor (8.1) the two characteristic plane waves have, respectively, $E_z = 0$ and $H_z = 0$, no matter what their common direction of propagation may be. They are therefore conveniently called TE (transverse electric) and TM (transverse magnetic), and any time-harmonic field formed by an angular spectrum of such waves is likewise the superposition of a TE field and a TM field. Explicitly, the field with suffix 1 in § 8.1.3, expressed in terms of M through (8.23), is a TE field; that with suffix 2, expressed in terms of Π through (8.24), is a TM field. Referring now to (8.30), (8.31) and (8.32), there is a forcible suggestion that the TE and TM fields can each be related to corresponding vacuum fields by respective simple scaling procedures. It is proposed to examine the general basis of this assertion by reference to Maxwell's equations.

Suppose that the vacuum field of a given current density is expressed as the superposition of TE and TM fields with respective current densities $\mathbf{J}_e$ and $\mathbf{J}_m$. Then the field of the given current density in the uniaxial medium with dielectric tensor (8.1) can be obtained by considering in turn the fields of $\mathbf{J}_e$ and $\mathbf{J}_m$.

When E_z is everywhere zero, Maxwell's equations for the uniaxial medium are identical with those for an isotropic dielectric with dielectric constant $\varkappa$. Plainly, therefore, the field of $\mathbf{J}_e$ in the uniaxial medium is obtained from the vacuum field of $\mathbf{J}_e$ merely by replacing ε_0 by $\varepsilon_0\varkappa$.

When H_z is everywhere zero the situation is not quite so trivial. But by writing down Maxwell's equations it is a straightforward matter to confirm the following statement: if $\mathbf{E}^0$, $\mathbf{H}^0$, $\mathbf{J}^0$ are the vectors of a TM vacuum field ($H_z^0 = 0$), then

$$\mathbf{E}(\mathbf{r}) = \left[\frac{1}{\varkappa}E_x^0(\mathbf{R}),\ \frac{1}{\varkappa}E_y^0(\mathbf{R}),\ \frac{1}{\sqrt{\varkappa\varkappa'}}E_z^0(\mathbf{R})\right], \tag{8.69}$$

$$\mathbf{H}(\mathbf{r}) = \left[\frac{1}{\sqrt{\varkappa}}H_x^0(\mathbf{R}),\ \frac{1}{\sqrt{\varkappa}}H_y^0(\mathbf{R}),\ 0\right], \tag{8.70}$$

$$\mathbf{J}(\mathbf{r}) = \left[J_x^0(\mathbf{R}),\ J_y^0(\mathbf{R}),\ \sqrt{\varkappa'/\varkappa}\,J_z^0(\mathbf{R})\right], \tag{8.71}$$

where

$$\mathbf{r} = (x, y, z), \quad \mathbf{R} = \left(\sqrt{\varkappa'}x,\ \sqrt{\varkappa'}y,\ \sqrt{\varkappa}z\right), \tag{8.72}$$

are the vectors of a TM field in the uniaxial medium. The TM field in the uniaxial medium due to the given current density $\mathbf{J}_m(\mathbf{r})$ can therefore be obtained by scaling, in the manner indicated, the vacuum field due to current density

$$\left[J_{mx}(\mathbf{R}'),\ J_{my}(\mathbf{R}'),\ \sqrt{\varkappa/\varkappa'}\,J_{mz}(\mathbf{R}')\right], \tag{8.73}$$

where

$$\mathbf{R}' = \left(\frac{x}{\sqrt{\varkappa'}},\ \frac{y}{\sqrt{\varkappa'}},\ \frac{z}{\sqrt{\varkappa}}\right). \tag{8.74}$$

For an electric dipole parallel to the z-axis the entire field is TM,

$$\mathbf{J}_m(\mathbf{r}) = (0, 0, p)\,\delta(x)\,\delta(y)\,\delta(z),$$

and (8.73) is simply $\varkappa\sqrt{\varkappa'}\mathbf{J}_m(\mathbf{r})$. The field in the uniaxial medium can be written down by appropriately scaling the vacuum dipole field, and the results (8.57)–(8.60) confirmed.

Where the current distribution is a surface density in the plane $z = 0$, as in § 8.1.2,

$$\mathbf{J}_m(\mathbf{r}) = [j_{mx}(x, y),\ j_{my}(x, y),\ 0]\,\delta(z).$$

Hence (8.73) is

$$\sqrt{\varkappa}\left[j_{mx}\left(\frac{x}{\sqrt{\varkappa'}}, \frac{y}{\sqrt{\varkappa'}}\right), j_{my}\left(\frac{x}{\sqrt{\varkappa'}}, \frac{y}{\sqrt{\varkappa'}}\right), 0\right]\delta(z),$$

so that anisotropy is not directly evident in the scaling of the original surface current density.

In the present statement of the problem the burden of the analysis devolves on the resolution of the vacuum field of a given current density into TE and TM fields. This is apparent, for example, in the case of the dipole perpendicular to the z-axis. One way of effecting the resolution is to introduce the plane wave spectrum representation of the vacuum field, and then write each plane wave as the superposition of a TE and a TM plane wave. This procedure leads to sensibly the same analysis as that which starts directly from the plane wave spectrum of the field in the uniaxial medium.

8.2. MAGNETO-IONIC MEDIUM

8.2.1. Surface Currents in Plane Normal to Magnetostatic Field

For the description of time-harmonic fields in a plasma with an imposed magnetostatic field the dielectric tensor takes the form

$$\mathscr{K} = \begin{pmatrix} \varkappa & -i\varkappa'' & 0 \\ i\varkappa'' & \varkappa & 0 \\ 0 & 0 & \varkappa' \end{pmatrix}, \tag{8.75}$$

where $\varkappa$, $\varkappa'$ and $\varkappa''$ are real if the medium is lossless. The z-axis is taken along the magnetostatic field.

For plane waves with space factor (8.5) the determinantal equation (2.38), when transformed by the operations described after (2.40), is

$$\begin{vmatrix} \varkappa - \mu^2 & -i\varkappa'' & l(\varkappa - \varkappa') - im\varkappa'' \\ i\varkappa'' & \varkappa - \mu^2 & m(\varkappa - \varkappa') + il\varkappa'' \\ l & m & \varkappa' \end{vmatrix} = 0,$$

which yields

$$\varkappa'(\mu^2 - \varkappa)^2 + (\varkappa - \varkappa')(l^2 + m^2)(\mu^2 - \varkappa) + (l^2 + m^2 - \varkappa')\varkappa''^2 = 0. \tag{8.76}$$

This quadratic equation for $\mu^2 - \varkappa$ gives two values for μ^2 (or n^2) as a function of l and m.

On the simplest form of magneto-ionic theory

$$\varkappa = 1 - \frac{X}{1 - Y^2}, \quad \varkappa' = 1 - X, \quad \varkappa'' = \frac{XY}{1 - Y^2}, \tag{8.77}$$

where

$$X = \omega_p^2/\omega^2, \quad Y = \Omega/\omega,$$

Ω being the angular frequency (eB_0/m) of free gyration of an electron in the magnetostatic field. Then with

$$\Delta = \frac{1 - Y^2}{X}(\mu^2 - \varkappa), \tag{8.78}$$

$$\varepsilon = \frac{Y^2}{1 - X}, \tag{8.79}$$

equation (8.76) becomes

$$\Delta^2 - \varepsilon(l^2 + m^2)\Delta + \varepsilon(l^2 + m^2 - 1 + X) = 0.$$

The solution is

$$\Delta = 1 - \nu \pm \sqrt{\nu^2 + Y^2 - 1},$$

where

$$\nu = 1 - \tfrac{1}{2}\varepsilon(l^2 + m^2);$$

and if this is written

$$\Delta = 1 + \frac{Y^2 - 1}{\nu \pm \sqrt{\nu^2 + Y^2 - 1}}$$

it gives

$$\mu^2 = 1 - \frac{X}{1 - \frac{1}{2}\varepsilon(l^2 + m^2) \pm \sqrt{\frac{1}{4}\varepsilon^2(l^2 + m^2)^2 - \varepsilon(l^2 + m^2) + Y^2}}. \tag{8.80}$$

Consider, now, the plane wave spectrum representation of the field of a surface current density in $z = 0$. It is convenient to introduce the inverse of (8.75), say

$$\mathscr{K}^{-1} = \begin{pmatrix} \alpha & i\gamma & 0 \\ -i\gamma & \alpha & 0 \\ 0 & 0 & \beta \end{pmatrix}. \tag{8.81}$$

Then if the expression (8.12) is prescribed for the magnetic field, the associated electric field can be written

$$\mathbf{E} = Z_0 \sum_{i=1,2} \Bigg[\pm \frac{\beta lmP_i^\pm + (1 - \beta l^2) Q_i^\pm}{n_i},$$
$$\mp \frac{(1 - \beta m^2) P_i^\pm + \beta lm Q_i^\pm}{n_i}, \; -\beta(mP_i^\pm - lQ_i^\pm)\Bigg] e^{ik_0(lx + my \mp n_i z)}, \tag{8.82}$$

in the derivation of which, first E_z is obtained from the z-component of $\mathscr{K}^{-1}\text{curl}\,\mathbf{H} = ik_0 Y_0 \mathbf{E}$, and subsequently E_x, E_y from the x and y components of curl $\mathbf{E} = -ik_0 Z_0 \mathbf{H}$. Furthermore, the P_i, Q_i relations can be put in the form

$$[l(\alpha\mu_i^2 - 1) - i\gamma mn_i^2]\, P_i^{\pm} + [m(\alpha\mu_i^2 - 1) + i\gamma ln_i^2]\, Q_i^{\pm} = 0, \qquad (i = 1,2), \tag{8.83}$$

a result which follows from a combination of the x- and y-components of $\mathscr{K}^{-1}$ curl $\mathbf{H} = ik_0 Y_0 \mathbf{E}$. Therefore the continuity of E_x and E_y across $z = 0$ again requires (8.16) to hold, and the representation (2.123), (2.124) is applicable.

When $\gamma = 0$, (8.83) reduces either to (8.15) or to $\mu_1^2 = 1/\alpha = \varkappa$, and it can be verified that the analysis of § 8.1.2 is recovered.

8.2.2. Surface Currents in Plane Parallel to Magnetostatic Field

In order to retain $z = 0$ as the plane of the surface currents the dielectric tensor is now taken to be

$$\mathscr{K} = \begin{pmatrix} \varkappa' & 0 & 0 \\ 0 & \varkappa & -i\varkappa'' \\ 0 & i\varkappa'' & \varkappa \end{pmatrix}, \tag{8.84}$$

with inverse

$$\mathscr{K}^{-1} = \begin{pmatrix} \beta & 0 & 0 \\ 0 & \alpha & i\gamma \\ 0 & -i\gamma & \alpha \end{pmatrix}. \tag{8.85}$$

The equation which replaces (8.76) is evidently

$$\varkappa'(\mu^2 - \varkappa)^2 + (\varkappa - \varkappa')(m^2 + n^2)(\mu^2 - \varkappa) + (m^2 + n^2 - \varkappa') \times \varkappa''^2 = 0,$$

and with $m^2 + n^2 = \mu^2 - l^2$ this takes the form

$$\varkappa(\mu^2 - \varkappa)^2 + [(\varkappa - \varkappa')(\varkappa - l^2) + \varkappa''^2](\mu^2 - \varkappa) + (\varkappa - \varkappa' - l^2) \times \varkappa''^2 = 0, \tag{8.86}$$

again giving two values of μ^2. The dependence of μ on l only is, of course, a consequence of the axial symmetry of the medium.

In the case (8.77), equation (8.86) is

$$\Delta^2 + 2\left[\frac{\varepsilon X}{1 - Y^2} - \tfrac{1}{2}\varepsilon(1 - l^2)\right]\Delta - \frac{\varepsilon(1 - \varepsilon)}{1 - Y^2}\left(l^2 + \frac{XY^2}{1 - Y^2}\right) = 0, \tag{8.87}$$

where now

$$\Delta = \frac{1-\varepsilon}{X}(\mu^2 - \varkappa) = \frac{1-\varepsilon}{X}(\mu^2 - 1) + \frac{1-\varepsilon}{1-Y^2}, \tag{8.88}$$

and ε is still given by (8.79). The solution is

$$\Delta = -\nu + \frac{1-\varepsilon}{1-Y^2} \pm \sqrt{\nu^2 + \varepsilon - 1},$$

where now

$$\nu = 1 - \tfrac{1}{2}\varepsilon(1 - l^2).$$

Thus

$$\frac{1-\varepsilon}{X}(\mu^2 - 1) = -\frac{1-\varepsilon}{\nu \pm \sqrt{\nu^2 + \varepsilon - 1}};$$

that is

$$\mu^2 = 1 - \frac{X}{1 - \tfrac{1}{2}\varepsilon(1-l^2) \pm \sqrt{\tfrac{1}{4}\varepsilon^2(1-l^2)^2 + \varepsilon l^2}}. \tag{8.89}$$

In the present case the magnetic field (8.12) has associated electric field

$$\mathbf{E} = Z_0 \sum_{i=1,2} \left\{ \pm\beta \frac{lmP_i^\pm + (\mu_i^2 - l^2)\, Q_i^\pm}{n_i}, \right.$$
$$-\left[\pm\frac{\alpha}{n_i}(\mu_i^2 - m^2) + i\gamma m\right] P_i^\pm - l\left(\pm\alpha\frac{m}{n_i} - i\gamma\right) Q_i^\pm,$$
$$\left. \left[\pm i\frac{\gamma}{n_i}(\mu_i^2 - m^2) - \alpha m\right] P_i^\pm + l\left(\pm i\gamma\frac{m}{n_i} + \alpha\right) Q_i^\pm \right\}$$
$$\times\, e^{ik_0(lx + my \mp n_i z)}, \tag{8.90}$$

this form being obtained directly from $\mathscr{K}^{-1}$ curl $\mathbf{H} = ik_0 Y_0 \mathbf{E}$. The P_i, Q_i relations, taken from the x-component of curl $\mathbf{E} = -ik_0 Z_0 \mathbf{H}$, are

$$\left(\alpha\mu_i^2 \mp i\gamma\frac{l^2 m}{n_i} - 1\right) P_i^\pm = \pm\, i\gamma\frac{l}{n_i}(\mu_i^2 - l^2)\, Q_i^\pm. \tag{8.91}$$

Evidently the continuity of E_x and E_y across $z = 0$ by no means implies (8.16), and the plane wave spectrum representation must be set up by the procedure outlined at the end of § 2.2.6.

At this stage it is probably simplest to substitute for $Q_i^\pm$ in (8.12) and (8.90) the multiple of $P_i^\pm$ given by (8.91). After a little algebra this gives

$$\mathbf{H} = \sum_{i=1,2} P_i^\pm \left\{1, \frac{\pm n_i(\alpha\mu_i^2 - 1) - i\gamma l^2 m}{i\gamma l(\mu_i^2 - l^2)}, \frac{m(\alpha\mu_i^2 - 1) \pm i\gamma l^2 n_i}{i\gamma l(\mu_i^2 - l^2)}\right\}$$
$$\times\, e^{ik_0(lx + my \mp n_i z)}, \tag{8.92}$$

and

$$\mathbf{E} = \frac{Z_0}{i\gamma} \sum_{i=1,2} P_i^{\pm} \left\{ \frac{\beta}{l}(\alpha\mu_i^2 - 1), -\frac{m[(\alpha^2-\gamma^2)\mu_i^2 - \alpha] \pm i\gamma n_i}{\mu_i^2 - l^2}, \right.$$
$$\left. \frac{\pm n_i[(\alpha^2-\gamma^2)\mu_i^2 - \alpha] - i\gamma m}{\mu_i^2 - l^2} \right\} e^{ik_0(lx+my\mp n_i z)}. \tag{8.93}$$

The continuity of E_x across $z = 0$ requires

$$\sum_{i=1,2} (\alpha\mu_i^2 - 1) D_i = 0, \tag{8.94}$$

that of E_y,

$$\sum_{i=1,2} \frac{1}{\mu_i^2 - l^2} \{m[(\alpha^2-\gamma^2)\mu_i^2 - \alpha] D_i + i\gamma n_i A_i\} = 0, \tag{8.95}$$

where

$$A_i = P_i^+ + P_i^-, \quad D_i = P_i^+ - P_i^-. \qquad (i = 1, 2). \tag{8.96}$$

And the match of the discontinuities in H_x and H_y to the surface current density, say

$$(j_x, j_y)\, e^{ik_0(lx+my)}, \tag{8.97}$$

requires

$$D_1 + D_2 = j_y, \tag{8.98}$$

$$\sum_{i=1,2} \frac{1}{\mu_i^2 - l^2} \left[\frac{n_i(\alpha\mu_i^2 - 1)}{i\gamma l} A_i - lmD_i \right] = -j_x. \tag{8.99}$$

8.2.3. Point Charge in Uniform Motion Parallel to Magnetostatic Field

The integrals in §§ 8.2.1, 8.2.2 are complicated, and even for dipole sources have not been evaluated explicitly. However, the radiation field can be found by the standard method of stationary phase. The phase

$$ik_0(lx + my \mp nz) \tag{8.100}$$

is stationary with respect to variations of l and m when

$$x \mp \frac{\partial n}{\partial l} z = 0, \quad y \mp \frac{\partial n}{\partial m} z = 0, \tag{8.101}$$

and the problem in essence reduces to an examination of these equations.

As a simple illustration, when

$$n^2 = \varkappa - (\varkappa/\varkappa')(l^2 + m^2)$$

the solution of (8.101) is easily seen to be

$$l = -\frac{\varkappa' x}{R}, \quad m = -\frac{\varkappa' y}{R}, \quad n = \pm\frac{\varkappa z}{R},$$

where

$$R^2 = \varkappa'(x^2 + y^2) + \varkappa z^2,$$

as in (8.39). The phase (8.100) is then just $-ik_0R$.

For the functions n^2 appearing in §§ 8.2.1, 8.2.2 the investigation of (8.101) is quite elaborate, and the results depend materially on the values of the parameters $\varkappa$, $\varkappa'$ and $\varkappa''$. The development is not pursued here.

The problem of Cerenkov radiation in a magneto-ionic medium is of considerable interest. This was discussed in § 8.1.6 for the special case of infinite magnetostatic field, and remains comparatively tractable because of the delta function behaviour of the frequency components of the plane wave spectrum functions.

Suppose the position of the point charge e at time t is $(vt, 0, 0)$, and that the frequency components of the field it generates are represented by (8.92), (8.93) (with affixes ω where appropriate) integrated over all real values of l and m. Then the right-hand sides of (8.98) and (8.99) must be identified thus,

$$j_x = \frac{e\omega}{4\pi^2 c}\,\delta(l + c/v), \quad j_y = 0, \tag{8.102}$$

these being the space-harmonic components of the frequency components of the current density associated with the moving point charge (see § 7.2.1). Hence, from (8.98),

$$D_1^\omega + D_2^\omega = 0, \tag{8.103}$$

which with (8.94) implies

$$D_1^\omega = D_2^\omega = 0; \tag{8.104}$$

that is

$$P_i^{+\omega} = P_i^{-\omega} \qquad (i = 1, 2). \tag{8.105}$$

Then $A_i^\omega = 2P_i^\omega$ (dropping the $\pm$ affixes), and the remaining equations, (8.95) and (8.99), give

$$(P_1^\omega, P_2^\omega) = -\frac{e}{8\pi^2 c}\,i\omega\,\frac{\gamma}{\alpha}\,\frac{l}{\mu_1^2 - \mu_2^2}\left(\frac{\mu_1^2 - l^2}{n_1}, \; -\frac{\mu_2^2 - l^2}{n_2}\right)\delta(l + c/v). \tag{8.106}$$

To get the frequency components of the field vectors the expressions (8.106) are substituted into (8.92), (8.93), and integrations taken over l and m. The l integration is trivial because of the delta function, and it is found that E_x^ω, from which the rate of radiation can be calculated, appears as

$$E_x^\omega = -\frac{Z_0 e}{8\pi^2 c}\frac{\beta\omega}{\alpha}e^{-i\omega x/v}\left\{\frac{(\alpha\mu_1'^2-1)(\mu_1'^2-c^2/v^2)}{\mu_1'^2-\mu_2'^2}\int_{-\infty}^{\infty}\frac{1}{n_1'}\right.$$

$$\times\, e^{ik_0(my\mp n_1'z)}\,dm+\frac{(\alpha\mu_2'^2-1)(\mu_2'^2-c^2/v^2)}{\mu_2'^2-\mu_1'^2}\int_{-\infty}^{\infty}\frac{1}{n_2'}$$

$$\left.\times\, e^{ik_0(my\mp n_2'z)}\,dm\right\}. \tag{8.107}$$

In (8.107) the dashes indicate that μ_i and n_i $(i = 1, 2)$ are evaluated at $l = -c/v$. It is recalled that μ_1^2, μ_2^2 are independent of m; for a conventional magneto-ionic medium they are given by (8.89).

The energy U radiated per unit length of the particle's track is now obtained by identifying it with $-eE_x$ evaluated at $(vt, 0, 0)$. Thus, from (7.60) and (8.107),

$$U = \frac{Z_0 e^2}{4\pi c}\left\{\left|\int\frac{\beta\omega}{\alpha}\,\frac{(\alpha\mu_1'^2-1)(\mu_1'^2-c^2/v^2)}{\mu_1'^2-\mu_2'^2}\,d\omega\right|\right.$$

$$\left.+\left|\int\frac{\beta\omega}{\alpha}\,\frac{(\alpha\mu_2'^2-1)(\mu_2'^2-c^2/v^2)}{\mu_2'^2-\mu_1'^2}\,d\omega\right|\right\}. \tag{8.108}$$

In this expression the first and second integrals are carried over only those positive ranges of ω for which $\mu_1' > c/v$ and $\mu_2' > c/v$, respectively: for if $\mu_1'^2(\mu_2'^2)$ is real, but less than c^2/v^2, then $n_1'(n_2')$ is pure imaginary; and if $\mu_1'^2$ is not real, then $\mu_2'^2 = \mu_1'^{2*}$, and the expression in curly brackets in (8.107) is pure imaginary when $y = z = 0$, because the second term is the negative conjugate complex of the first. The modulus signs in (8.108) are included to allow for the fact that the signs of the real values of n_1' and n_2' in (8.107) are not determined until it has been ascertained, in any particular case, whether the z-components of the directions of phase propagation and time-averaged energy flow associated with the plane wave of the spectrum have common or opposite signs.

The integrands in (8.108) represent, in effect, the frequency spectrum of the radiated power, and this in turn is intimately related to the directional characteristics of the radiation through the condition $\mu \cos\theta = c/v$. Here θ is the angle made with the magnetostatic field, and for each value of ω the condition assigns a value to θ.

In conclusion it is worth confirming that the results for the two special cases previously considered, in §§ 7.2.2 and 8.1.6, can be recovered from (8.108). In the former case the medium is isotropic, and

$$\gamma = 0, \quad \mu_1^2 = \mu_2^2 = 1/\beta = 1/\alpha;$$

moreover, with the modulus signs discarded and a common ω range of integration, (8.108) is simply

$$U = \frac{e^2 Z_0}{4\pi c} \int \frac{\beta\omega}{\alpha} \left[\alpha(\mu_1'^2 + \mu_2'^2) - \left(1 + \frac{\alpha c^2}{v^2}\right)\right] d\omega,$$

in evident agreement with (7.59). In the latter case

$$\gamma = 0, \quad \alpha = 1, \quad \mu_1^2 = 1, \quad \mu_2^2 = 1/\beta + (1 - 1/\beta)\, l^2;$$

hence

$$\beta(\mu_2'^2 - c^2/v^2) = 1 - c^2/v^2$$

and (8.108) leads to (8.68).

ANNOTATED BIBLIOGRAPHY

The two-dimensional plane wave spectrum of fields due to aperture antennas seems to originate with

Woodward, P. M. and Lawson, J. D. (1948). The theoretical precision with which an arbitrary radiation-pattern may be obtained from a source of finite size. *Proc. IEE,* **95III**, 363–70.

and

Booker, H. G. and Clemmow, P. C. (1950). The concept of an angular spectrum of plane waves, and its relation to that of polar diagram and aperture distribution. *Proc. IEE*, **97III**, 11–17.

The former paper deals with the approximate synthesis of specified radiation patterns by suitable distributions of field over either a given aperture or a given linear array. The latter paper links the concept of an angular plane wave spectrum with those of a polar diagram and the Fourier transform of an aperture field distribution, showing that the angular spectrum is more general; when introducting the concept of an angular spectrum of plane waves references is made to the earlier work

Ratcliffe, J. A. (1946). *Journal IEE*, **93III**(A), 28.

The technique of using an angular plane wave spectrum field representation was applied initially to study ionospheric scattering and diffraction. Of particular note are

Booker, H. G., Ratcliffe, J. R. and Shinn, D. H. (1950). Diffraction from an irregular screen with applications to ionospheric problems. *Phil. Trans. Roy. Soc. A*, **242**, 579–607,

and the review article

Ratcliffe, J. A. (1956). Some aspects of diffraction theory and their applications to the ionosphere. *Rep. Prog. Phys.*, **19**, 188–267.

In both of these the scattering from statistically rough surfaces is investigated. The plane wave spectrum representation method was extended to three-dimensional fields in

Brown, J. (1958). A theoretical analysis of some errors in aerial measurements. *Proc. IEE*, **105C**, 472–5.

The half-plane diffraction problem was first solved by

Sommerfeld, A. (1896). Mathematische theorie der diffraction. *Math. Ann.*, **47**, 317–74.

An account written in English is given in

Sommerfeld, A. (1954). *Optics*, 245–65. Academic Press, New York.

The curious reader will also be interested in the thorough review article

Bouwkamp, C. J. (1954). Diffraction theory. *Rep. Prog. Phys.*, **17**, 35–100,

and in Chapter XI, entitled 'Rigorous diffraction theory', which Phillip Clemmow wrote for the classic

Born, M. and Wolf, E. (1980). *Principles of optics* (6th edn). Pergamon Press, Oxford.

In this chapter Phillip Clemmow discusses the concepts of angular spectra of plane waves and the dual integral equation formulation of diffraction problems, and then goes on to solve the problems of both two- and three-dimensional diffraction of a plane wave by a half-plane. The half-plane diffraction of a line source field is also discussed, as are the problems of scattering from two parallel half-planes, an infinite stack of parallel, staggered half-planes, and from a strip. Phillip Clemmow also provided, in Appendix III of the book, a lucid presentation of the methods of stationary phase and steepest descent.

There is a vast literature connected with the so-called 'mixed path' problem and its applications. Clemmow's solution for a two-part plane surface was first presented in .

Clemmow, P. C. (1954). Radio propagation over a flat earth across a boundary separating two different media. *Phil. Trans. Roy. Soc. Lond. A*, **246**, 1–55.

This paper, a masterpiece of erudition, contains an extensive list of related publications, including the seminal one by Weyl, H. (1919) *Annal. Physik* (Leipzig), **60**, 481.

Alternative approaches to this problem may be found, the first one being

Feinberg, E. (1946). On the propagation of radio waves along an imperfect surface. *J. Phys. (Moscow)*, **10**, 410–40.

Applying the rationale of operational calculus, and starting from Green's integral theorem, we have also

Bremmer, H. (1954). The extension of Sommerfeld's formula for the propagation of radio waves over a flat earth, to different conductivities of the soil. *Physica's Grav.*, **20**, 441–60.

The paper

Wait, J. R. (1956). Mixed path ground wave propagation, 1. Short distances. *J. Res. Nat. Bur. Stand.*, **57**, 1–15.

makes nice use of the compensation theorem of Montcath to solve the problem. Clemmow's approach was taken in

Millar, R. F. (1967). Propagation of electromagnetic waves near a coastline on a flat earth. *Radio Sci.*, **2**(3), 261–86.

A different method of solving the mixed-path problem is given in

Walsh, J. and Donnelly, R. (1987). A new technique for studying propagation and scattering for mixed paths with discontinuities. *Proc. Roy. Soc. Lond. A*, **412**, 125–67.

The field of a moving point charge was traditionally solved by the introduction of intermediate vector and scalar potentials. The Liénard–Wiechert potentials are usually used to give the electric field for, say, an electron moving with constant velocity; an alternative approach in this case is to apply a Lorentz transformation to the fields of a static charge. This is discussed in

Panofsky, W. K. H. and Philips, M. (1962). *Classical electricity and magnetism* (2nd edn), §19.2. Addison Wesley, Reading, MA.

The field of an arbitrarily moving charge can also be found by the method of 'normal variables' and Fourier transformation, as is described in

Cohen-Tannoudji, C., Dupont-Roc, J. and Grynberg, G. (1989) *Photons and atoms*, pp. 66–8. Wiley Interscience, New York.

An approach that bypasses the use of potentials is given in

Donnelly, R. and Ziolkowski, R. W. (1994). Electromagnetic field generated by a moving point charge: a fields-only approach. *Am. J. Phys.*, **52**(10), 916–22.

Two excellent books where the angular plane wave spectrum representation method and its use in diffraction and antenna problems have been clearly presented are

Clarke, R. H. and Brown, J. (1980). *Diffraction theory and antennas.* Ellis Horwood Ltd, Chichester.

and

Jull, E. V. (1981). *Aperture antennas and diffraction theory.* Peter Peregrinus Ltd, Stevenage.

Rod Donnelly
Memorial University of Newfoundland

INDEX